Springer Theses

Recognizing Outstanding Ph.D. Research

Aims and Scope

The series "Springer Theses" brings together a selection of the very best Ph.D. theses from around the world and across the physical sciences. Nominated and endorsed by two recognized specialists, each published volume has been selected for its scientific excellence and the high impact of its contents for the pertinent field of research. For greater accessibility to non-specialists, the published versions include an extended introduction, as well as a foreword by the student's supervisor explaining the special relevance of the work for the field. As a whole, the series will provide a valuable resource both for newcomers to the research fields described, and for other scientists seeking detailed background information on special questions. Finally, it provides an accredited documentation of the valuable contributions made by today's younger generation of scientists.

Theses are accepted into the series by invited nomination only and must fulfill all of the following criteria

- They must be written in good English.
- The topic should fall within the confines of Chemistry, Physics, Earth Sciences, Engineering and related interdisciplinary fields such as Materials, Nanoscience, Chemical Engineering, Complex Systems and Biophysics.
- The work reported in the thesis must represent a significant scientific advance.
- If the thesis includes previously published material, permission to reproduce this must be gained from the respective copyright holder.
- They must have been examined and passed during the 12 months prior to nomination.
- Each thesis should include a foreword by the supervisor outlining the significance of its content.
- The theses should have a clearly defined structure including an introduction accessible to scientists not expert in that particular field.

More information about this series at http://www.springer.com/series/8790

Christina Giarmatzi

Rethinking Causality in Quantum Mechanics

Doctoral Thesis accepted by
The University of Queensland, Brisbane,
Australia

 Springer

Author
Dr. Christina Giarmatzi
School of Mathematics and Physics
The University of Queensland
Brisbane, QLD, Australia

Supervisor
Prof. Andrew White
School of Mathematics and Physics
The University of Queensland
Brisbane, QLD, Australia

ISSN 2190-5053   ISSN 2190-5061   (electronic)
Springer Theses
ISBN 978-3-030-31932-8   ISBN 978-3-030-31930-4   (eBook)
https://doi.org/10.1007/978-3-030-31930-4

This Springer imprint is published by the registered company Springer Nature Switzerland AG
The registered company address is: Gewerbestrasse 11, 6330 Cham, Switzerland

*To the two friends, who have sustained me
over the years, Zoi and Petros
To a beautiful soul, Stephen
And to a shining star, Ben*

Supervisor's Foreword

Discussions of cause and effect are as old as human history, with the concept holding great sway from Greek philosophy through to Buddhism. In science, consideration of causation has been fundamentally influential in fields ranging from physics to epidemiology. Famously, the fundamental role of causality in quantum mechanics was teased into clear view by the decades-long study of Bell's inequality; most recently, causal networks and discovery algorithms have had important applications in machine learning, which is now interpolating every aspect of our daily lives. So, gentle reader, you may be justified in wondering what more can there be possibly to say on the topic?

In this thesis, Christina introduces and applies a new definition of causal process, one which applies equally to classical and quantum theory, and indeed to any future post-quantum theories. She introduces the theory-independent notion that the cause and effect dependencies between events in a process can depend on other events within the process. This original and striking insight has great power, providing a new tool for consideration of quantum gravity, as well as having applications in quantum information processing, including computation and communication.

Christina's assessors remarked that her thesis "…is a substantial contribution to the knowledge in the field of causality", that it "demonstrates remarkable depth of research and breadth of knowledge", and that Christina has made "a contribution over and above what one would normally expect from a Ph.D. student … [she] has contributed broadly to multiple facets of the field of quantum causality, rather than narrowly focusing on one or two problems." Christina's thesis is indeed all of these things, and it is also eminently readable, endlessly fascinating, and a must for anyone interested in foundational questions about the universe we live in.

So waste no more time on this foreword but turn the page and begin your journey, you will find—as I did—that you are in for a treat.

Brisbane, Australia   Prof. Andrew White
May 2019

Abstract

Causality has always believed to be a well-defined fundamental concept. However, the late efforts for a theory of quantum gravity have suggested that causal structures may not only be dynamical—as in general relativity—but also indefinite; the way a quantum observable has an indefinite value prior to its measurement. Based on this observation, it was proposed that a good way to study causality might be through a probabilistic framework for correlations between events without an inherent assumption of a definite causal structure. This way, we are free to formulate causality in terms of correlations between events, and later we can see what happens when the correlations are obtained from quantum events.

In this thesis, we developed such a general probabilistic framework, which is independent of the theory that describes the events. We proposed a concept of causality, taking into account that the causal order may be dynamical; an event may affect the causal order of other events in its future. In that theory-independent framework, we formulated causality in terms of constraints on the correlations. In the case where the events are described by quantum mechanics, there is a special way to describe correlations: through the process matrix. In that theory-dependent framework, we found that causality is expressed in terms of simple conditions on the process matrix. We observed the fascinating differences in which causality manifests at the level of the two frameworks. We worked further on the latter one, to develop mathematical tools to computationally and experimentally test situations incompatible with a definite causal order. We also developed computational methods to obtain restrictions of causality in terms of inequalities, for a given scenario. We finished our study with a powerful and promising field of causality: quantum causal discovery. Assuming that there is a well-defined and fixed causal order between a number of events, causal discovery aims at inferring the causal structure through data from the events. First, we performed two complementary experiments that rule out a class of classical hidden-variable causal models for Bell correlations. Finally, abandoning the idea that quantum events have a classical causal model, we use a quantum causal modeling framework to write the first quantum causal discovery algorithm.

Acknowledgements

I acknowledge the traditional owners of the land on which the University of Queensland is situated, the Turrbal and Jagera people.

I would like to begin by expressing my gratitude to the person that gave me the opportunity to do my Ph.D. in this wonderful place: we all know him, we all love to see him, and we always know he is near, because his laughter echoes throughout the Physics building: it is Andrew White. Thank you Andrew for the opportunity, for welcoming me to the team, and for all the great advice. I am also very grateful to the rest of my advisory team: Gerard Milburn for all the inspiring conversations about physics, and Fabio Costa for your patience, daily teachings, and encouragement which helped me to become a better researcher.

Now I have to take a step back and thank my family, for providing me with a home and with values that my dad still repeats: 'wisdom is the greatest gift'; and, 'in every minute that passes, one should learn'. Thank you for understanding that I have to be on the other side of the world, and for still loving me. I also thank my cousin Apostolis, for always being a great support to me, and for giving me the book Chaos, by James Gleick, which led me to physics.

A very special thanks to my friends, Zoi and Petros; your acceptance, encouragement, and innumerable moments of laughter got me through the best of times and the worst of times. I thank Isa for being a good sister here in Brisbane, for the wisdom and silliness and laughters we shared. I thank my office mates, for all the fun times at the office and all those pats on the shoulder that kept me going. I thank Stephen, for being my friend and my teacher, and for his lessons of courage, bravery, and kindness. Finally, I thank Ben, for all the wonderful times, amazing guitar solos, and the continuing adventures we share.

Contents

Chapter 1
Introduction

1.1 What Happened to Causality

From the beginning of scientific reasoning the concept of causality has been a fundamental concept with which we understand the mechanisms of nature: A is the cause of B, and there is nothing more to it. General relativity uses this concept to establish causal structures: the set of possible causal relations between any possible event localised in spacetime. It teaches us that causal structures can be dynamical because massive objects change the causal structure around them, giving an overall very satisfying theory on how the world works.

This clear understanding of causality, a profound pillar of physics, has recently started to get blurry; causing a real stir on the idea that we had figured things out. It mainly started with the effort to bring together two contradicting theories: general relativity and quantum mechanics. Briefly, quantum mechanics is a probabilistic theory that assumes a fixed causal structure, i.e. quantum mechanics is the theory to describe what small objects are doing on a fixed spacetime shaped by a massive object that is not going anywhere near the realm of uncertainty. General relativity on the other hand, is a deterministic theory that describes what is happening to objects that are so massive that they shape space-time around them and consequently the motion of other massive objects. One effort to bring these theories together was developed by Hardy [1, 2] who proposed a scheme for quantum gravity, in 2008. His idea was that such a scheme should be probabilistic in nature, like quantum mechanics, but with a dynamical—not fixed—causal structure, like in general relativity. This led to the idea that causal structures could be probabilistic and 'indefinite', in the same sense that we cannot ascribe a value to an unobserved quantum variable.

Soon after Hardy's proposal, interest began to grow in the causal structure of quantum events and what can be done in terms of quantum computation. In 2009, Chiribella et al. [3] proposed a novel architecture for quantum circuits beyond the standard quantum circuit model. While in the latter quantum operations occur one after the other in a fixed time-sequence, in their proposed architecture the wires that

© Springer Nature Switzerland AG 2019
C. Giarmatzi, *Rethinking Causality in Quantum Mechanics*, Springer Theses,
https://doi.org/10.1007/978-3-030-31930-4_1

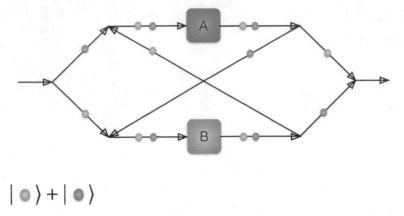

$$|\,\bullet\,\rangle + |\,\bullet\,\rangle$$

Fig. 1.1 The quantum switch: a quantum circuit with movable wires connecting the operations. Depending on the input state, the system can take different paths, leading to different orders for A and B

connect the operations can be in superposition of different connections, allowing for some sort of superposition of different quantum circuits.

Although this seems like a rather normal concept—a photon and a beamsplitter is all it takes to create superposition of the path of the photon—its implications were revolutionary. Superposition of the path of the photon is somehow understood; but when along each path other events occur in a particular order, we are led to superposition of the order of events. For two events, such a scheme can be represented as shown in Fig. 1.1, which is the famous *quantum switch* [3].

At the same time, Chiribella, D'Ariano and Perinotti developed the framework of *quantum combs* [4], which can describe situations like the quantum switch as a map (which they call a *supermap*). The first result on this weird quantum circuit is that it offers a computational advantage [5]: Imagine that you are promised that the operations on A and B are unitaries that either commute or anticommute, and that you have to guess which one it is by putting them in a quantum circuit and use a single query. A reasoning that involves results on state discrimination proves that a quantum circuit with a fixed causal order cannot discriminate between commuting versus non-commuting unitaries with one use of the circuit. As it turns out, the quantum switch can very well do that, with a single query.

Although this first computational advantage of the quantum switch promised that many interesting things can happen using superposition of circuits, it also raised a lot of questions: what does it mean to have a superposition of 'A causes B' and 'B causes A'? Can A signal to B *and* B signal to A? What kind of causal structure would allow that? And what are the massive objects doing to create such weird causal structure?

It was clear that these cases cannot be treated with the traditional quantum mechanics framework in which causality is always definite; there needed to be one that allows for quantum mechanics locally but also for a non-fixed causal structure.

Although the first step was done by the framework of quantum combs, a more general framework was developed by Oreshkov et al. [6], in 2012. It is a framework for the study of correlations between quantum events without the assumption of a predefined and definite causal order between them. Through these correlations, we can then talk about their causal structure. Using this formalism, it was shown that it is possible to obtain correlations between events that are incompatible with an underlying causal order. The environment of the events is represented by a mathematical object called *process matrix*. A process matrix is the resource that allows the events to be correlated in a particular way, compatible or not with a definite causal structure. Most of the chapters of this thesis are using this formalism to explore the notion of causality and build a few tools to study it (Chaps. 2–4).

The process matrix formalism has also served to build a framework for causal modeling [7] and causal discovery [8] in the quantum regime. In the classical case, Pearl [9], in 2000, developed the essential tools for causal modeling and causal inference: how to discover the causal model from correlations. Note that a causal model is a graph where nodes represent variables and directed arrows representing causal influences. However, these tools do not apply in the quantum case as shown with the famous example of a Bell-type experiment: Bell-correlations do not admit a causal model that obeys the well-justified Bell's assumptions. This means that there is something wrong with the idea that quantum correlations should fit in the classical causal model picture. Although the efforts continue to recover a classical causal model for quantum correlations (Chap. 5), this suggests that we may need to abandon the idea of classical causal models for quantum events. This was done by Costa and Shrapnel [7], in 2016, using the process matrix formalism (and within a different framework, independently by Allen et al. [10] in the same year). This led to the first causal discovery algorithm for arbitrary (but fixed) causal structures in the quantum regime [8] (Chap. 6).

The development of the process matrix formalism inspired a lot of research which transformed into a field some call 'quantum causality'. Some results—before, during and after the submission of this thesis—involve computational advantages of indefinite causal order [11–17], a general study of causality for quantum events [18–26], and even experimental realisations of the quantum switch [27–31]. The field is still growing with the promise of novel routes on how to make sense of causality in the quantum regime and more experimental realisations of indefinite causal structures.

1.2 Short Description of the Chapters

Different chapters of this thesis fit in a different way in this study of causality. Chapter 2, based on [18], is devoted to the notion of causality itself from two different perspectives. First, from a theory-independent perspective, we study how causality is expressed between a number of local experiments, without assuming any particular theory that describes the experiments. We formulate the conditions that causality imposes on the possible correlations between the experiments, for n experiments.

We take into account the possibility of dynamical causal order: the fact that one event can influence the causal order of events in the future. Our second perspective is a theory-dependent one: the experiments are described by standard quantum mechanics and we use the process matrix formalism, which we extended to n parties as well. We found that, depending on whether we are at the theory-independent or theory-dependent level, their classification in terms of causality changes. We study this phenomenon and propose a classification of situations at both levels.

Chapter 3, based on [22, 31], is about developing a mathematical tool to use with the process matrix formalism to be able to recognize situations incompatible with causality. It is the analogue of an entanglement witness, but for causality, called a causal witness. A causal witness corresponds to a set of operations that need to be performed by the parties to prove that a process matrix—a description of the situation—is causally nonseparable, i.e. incompatible with a causal order between the experiments. We show that, in contrast with an entanglement witness, a causal witness can always be found, as it can be cast as a SemiDefinite Program (SDP)—a type of optimization algorithm with linear constraints—which can be solved efficiently. We present the SDPs we developed for the general bipartite case and a particular tripartite case. We also found the optimal causal witness for the quantum switch, a causally nonseparable process matrix with experimental realization. On a separate project, we plan to build our own quantum switch. We present how we can tailor our SDPs to match some experimental requirements, which we did for our own implementation, for which we report its progress.

Chapter 4, based on [32], is about studying the polytope of correlations between a number of parties that are compatible with causality: the causal polytope. This is the analogue of the local polytope: the set of probabilities arising from measurements on multipartite separable states; its facets correspond to Bell-type inequalities whose violation by a given set of measurements verifies entanglement of the state (that is also not Bell-local of course). We first prove that correlations of the outcomes of a number of parties, given their settings, when they satisfy causality, form a convex polytope. Given this result, the causal polytope for any number of parties and settings and outcomes for them can be characterized by its vertices. We characterized the simplest tripartite causal polytope: we obtained its facets and classified them in families of causal inequalities. Finally, using a generalized 'see-saw' approach we obtained process matrices and the operations that produce correlations that violate a given family of inequalities.

In Chap. 5, based on [33], we take a leap into the experimental world. We are interested in a particular classical (hidden variable) causal model compatible with correlations, arising from a Bell-type experiment. One classical (hidden variable) causal model that can explain these correlations involves a causal influence from the setting of one measurement station to the outcome of the other. We test, experimentally, this causal model by performing two complementary experiments: in one experiment we perform interventions on the output of one party, and observe the changes in the statistics on the outcome of the other party. We found that the change in the statistics was insufficient to explain the observed correlations, as calculated in Ref. [34]. As proposed in the same reference, we then implement a measurement scheme different to the traditional—it involves three settings for each party, as opposed to two. This is our second experiment and it allows for a violation of a new inequality [34], that rules out the class of causal models we are interested in. This method is device-independent (as opposed to the first one) which makes the claim even stronger. The result is that we ruled out a class of hidden variable causal models that explain the Bell correlations. In the next and final chapter, we give up the notion of having a classical causal model for variables arising from quantum systems.

In Chap. 6, based on [8], we use a causal modeling framework for quantum events [7], which arose from the process matrix formalism, and use it to write an algorithm for causal discovery (or causal inference). The causal models on which we are focusing are those for parties that perform quantum operations on their local laboratories that receive and send out quantum systems. We associate a causal link from party A to party B, if there is a quantum channel from the output system of A to the input system of B. Such a quantum channel appears in their process matrix, and corresponds to a linear constraint on the process matrix. Hence, given the process matrix for a number of parties, we subject it to all possible linear constraints (causal arrows). The set of the linear constraints that are satisfied by the process matrix defines a unique and minimal (there are no superfluous causal arrows) causal model for the parties. We write an algorithm in MatLab that first detects if all the relevant common causes are included in the process, a condition called Markovianity. For a Markovian process, the algorithm outputs a unique and minimal causal model; namely the causal relations and the corresponding mechanisms, represented as quantum states and channels. Our algorithm provides a first step towards more general methods for quantum causal discovery.

References

1. Hardy L (2007) Towards quantum gravity: a framework for probabilistic theories with non-fixed causal structure 40:3081–3099
2. Hardy L, Myrvold WC, Christian J (2009) Quantum gravity computers: on the theory of computation with indefinite causal structure. Springer, Dordrecht, pp 379–401
3. Chiribella G, D'Ariano GM, Perinotti P, Valiron B (2013) Quantum computations without definite causal structure. Phys Rev A 88:022318. arXiv:0912.0195
4. Chiribella G, D'Ariano GM, Perinotti P (2008) Transforming quantum operations: quantum supermaps 83:30004
5. Chiribella G (2012) Perfect discrimination of no-signalling channels via quantum superposition of causal structures. Phys Rev A 86:040301. arXiv:1109.5154
6. Oreshkov O, Costa F, Brukner Č (2012) Quantum correlations with no causal order. Nat Commun 3:1092
7. Costa F, Shrapnel S (2016) Quantum causal modelling. New J Phys 18:063032
8. Giarmatzi C, Costa F (2018) A quantum causal discovery algorithm. npj Quantum Inf 4:17
9. Pearl J (2009) Causality. Cambridge University Press
10. Allen J-MA, Barrett J, Horsman DC, Lee CM, Spekkens RW (2016) Quantum common causes and quantum causal models 1609:09487
11. Araújo M, Costa F, Brukner Č (2014) Computational advantage from quantum-controlled ordering of gates. Phys Rev Lett 113:250402. arXiv:1401.8127

12. Zhao X, Giulio C (2019) Advantage of indefinite causal order in quantum metrology. In: Quantum information and measurement (QIM) V: Quantum Technologies, F5A.23, Optical Society of America, Rome
13. Feix A, Araújo M, Brukner Č (2015) Quantum superposition of the order of parties as a communication resource. Phys Rev A 92:052326
14. Guérin PA, Feix A, Araújo M, Brukner Č (2016) Exponential communication complexity advantage from quantum superposition of the direction of communication. Phys Rev Lett 117:100502
15. Giulio C (2018) Indefinite causal order enables perfect quantum communication with zero capacity channel 1810:10457
16. Eblcr D, Salek S, Chiribella G (2018) Enhanced communication with the assistance of indefinite causal order. Phys Rev Lett 120:120502
17. Sina S (2018) Quantum communication in a superposition of causal orders 1809:06655
18. Oreshkov O, Giarmatzi C (2016) Causal and causally separable processes. New J Phys 18:093020
19. Oreshkov O (2018) Time-delocalized quantum subsystems and operations: on the existence of processes with indefinite causal structure in quantum mechanics 1801:07594
20. Oreshkov O, Cerf NJ (2015) Operational formulation of time reversal in quantum theory. Nat Phys 11:853 EP
21. Oreshkov O, Cerf NJ (2016) Operational quantum theory without predefined time 18:073037
22. Araújo M et al (2015) Witnessing causal nonseparability. New J Phys 17:102001
23. Milz S, Pollock FA, Le TP, Chiribella G, Modi K (2018) Entanglement, non-markovianity, and causal non-separability 20:033033
24. Giarmatzi C, Costa F (2018) Witnessing quantum memory in non-Markovian processes. arXiv:1811.03722 [quant-ph]
25. Ho CTM, Costa F, Giarmatzi C, Ralph TC (2018) Violation of a causal inequality in a spacetime with definite causal order. arXiv:1804.05498 [quant-ph]
26. Wechs J, Abbott AA, Branciard C (2019) On the definition and characterisation of multipartite causal (non)separability 21:013027
27. Friis N, Dunjko V, Dür W, Briegel HJ (2014) Implementing quantum control for unknown subroutines. Phys Rev A 89:030303
28. Procopio LM et al (2015) Experimental superposition of orders of quantum gates. Nat Commun 6:7913 EP
29. Friis N, Melnikov AA, Kirchmair G, Briegel HJ (2015) Coherent controlization using super-conducting qubits. Sci Rep 5:18036 EP
30. Rubino G et al (2017) Experimental verification of an indefinite causal order. Sci Adv 3:e1602589
31. Goswami K et al (2018) Indefinite causal order in a quantum switch. Phys Rev Lett 121:090503
32. Abbott AA, Giarmatzi C, Costa F, Branciard C (2016) Multipartite causal correlations: polytopes and inequalities. Phys Rev A 94:032131
33. Ringbauer M et al (2016) Experimental test of nonlocal causality. Sci Adv 2. http://advances.sciencemag.org/content/2/8/e1600162.full.pdf
34. Chaves R, Kueng R, Brask JB, Gross D (2015) Unifying framework for relaxations of the causal assumptions in bell's theorem. Phys Rev Lett 114:140403

Chapter 2
Causal and Causally Separable Processes

2.1 Back Story

This chapter is based on the long article in Ref. [1], which includes a video abstract produced by yours truly. The initial objective of this project was to extend the work presented in Ref. [2] into the multipartite case. That work was a mathematical framework, called the process matrix framework, which is a way to expresss correlations between a number of local experiments between an input and an output system, without any reference to their causal order—not even assuming that one exists. Examples were studied for the bipartite case, where they showed a scenario that yields correlations between the experiments that would defy causality. In order to make the latter claim, they defined what it means for two parties to obey causality.

Extending the mathematical framework for three parties was easy, but to define what it means that three (and eventually n) parties obey causality, turned out to be quite a task. Think of a situation respecting causality, for two parties. Surely a rigorous notion of causality is not given yet, but following our intuition we can say: only the past affects the future. Then we come up with these three cases: A before B, B before A and A and B being causally independent. Then the most general case would be a probabilistic mixture of these three cases. Indeed, a causal *process*—the collection of joint probabilities of the local outcomes of the experiments, given their settings—is one that can be written as a probabilistic mixture of processes compatible with these three cases, and the probability weights are independent of all events.

Now think of three parties and imagine what would be the most general situation compatible with our intuition about causality: all possible causal configurations in which some parties are before others and some parties are causally independent. But wait, this is not the only possible scenario. Imagine the parties A, B and C. It can be that C is first, and depending on the state of his output system, the system could take a route that goes first to A and then to B, or another route that goes B first and then to A. This is compatible with our intuition as a horizontally polarized photon would go through a polarizing beam splitter and a vertical one would make a 90-degree turn, and hence the photon would take a different spatial path. Keeping this in mind,

© Springer Nature Switzerland AG 2019

C. Giarmatzi, *Rethinking Causality in Quantum Mechanics*, Springer Theses,
https://doi.org/10.1007/978-3-030-31930-4_2

what would a causal process look like, for our three parties? In the bipartite case a causal process was a convex mixture of processes compatible with all possible causal configurations for the two parties, with the probability weights being independent of all relevant events. But for our three parties this is not true anymore. We cannot assign an independent probability weight for the process to turn out to be $C \prec A \prec B$ (meaning, C to A to B) or $C \prec B \prec A$ as that probability would *depend* on what Charlie does. Right there, we observed the need for a definition of causality that would be taken into account of such a 'dynamical' causal order. We could not rely on the obvious suggestion from the bipartite case that a causal situation is one that can be written as a mixture of scenarios compatible with a fixed causal order between the events with independent probability weights, because those probability weights have to be allowed to be *dependent* on the settings of some parties.

Hence, we searched for a new definition of causality, one that would allow the settings of the parties to affect the causal order of their future parties. This is the core result of this work. We saw how this definition manifests itself in the correlations of joint probabilities of the local outcomes of the experiments, given their settings, that is, the process framework, at a theory-independent level. We then applied the definition of causality within the quantum process framework (where the local experiments are described by quantum mechanics) and saw how it is expressed through simple conditions on the process matrix, in the tripartite case.

Furthermore, we explored the interplay between the two different levels of investigation: general process framework and quantum process framework: level of probabilities in theory-independent terms, vs level of process matrices where the local theory is quantum mechanics. Within the quantum process framework, we saw that a number of peculiar effects arise, which led to various kinds of definitions for different situations. For example, within that framework—in which we can describe situations using a process matrix—causality imposes constraints on these matrices. But it turns out that these constraints have a different effect than the constraint of causality on the probabilities of the local outcomes of the parties. For example, there can be a process matrix which does not obey the conditions of causality, however the probabilities that are obtained by the parties, could agree with causality. Or, an even more peculiar effect, a process matrix that agrees with causality—thus giving rise to probabilities also agreeing with causality—can give rise to probabilities not agreeing with causality, when the parties are supplied with extra entangled input systems.

Finally, we classified the different types of processes and process matrices that we found, depending on the conditions under which they respect causality. The differences between our two levels of investigation—process versus process matrix—will be clear later, but general idea is that compatibility with causal order differs whether we talk about a process matrix—a description of the *environment* that connects the parties (or their resource)—and the probabilities (process) that can arise from a process matrix. Certainly a process matrix that obeys causality cannot give rise to correlations that do not, but the opposite is not true: a process matrix that does not obey causality may or may not be able to produce probabilities that do. Something similar happens with entangled states: they are not separable, yet they do not necessarily yield probabilities that can violate any Bell-type inequalities.

It is understood that the reader might be confused or skeptical about our approach. Surely there must be an error in our definition of causality as its manifestations at different levels seem to disagree. But the problem lies elsewhere: our understanding is that the peculiarities of the process matrix description arise in the very core assumption of the process matrix framework: local quantum mechanics. Now the reader might not be so surprised as many peculiar effects arise when quantum mechanics is involved. We know that every property of a physical system cannot escape its quantum character, hence it is not well-defined prior to and independently of its measurement, and, it may be the case that the notion of causal order, although not a property of a single physical system but rather their environment, may not have escaped either.

2.2 Introduction

Dynamical and indefinite causal structures have recently attracted a lot of interest; first from a foundational point of view and later for the possibility of enhanced quantum information processing [2–27]. The foundational aspect concerns the long standing search for a theory of quantum gravity, where the causal structure is expected to be dynamical as in General Relativity but fundamentally probabilistic in nature, while the interest from quantum information processing comes from the search for novel quantum architectures beyond the standard circuit model. Regardless of the motivation, operational approaches to causal order in a probabilistic setting have provided a clearer view of causality as a classical and as a quantum concept, leaving us with a promise for technological applications.

Recently, a framework was proposed in Ref. [2] to investigate the role of causality in an operational setting. The aim of that framework was to study the correlations between local experiments without the assumption of a predefined causal order between them. It was found out that a definite causal structure in which the experiments are embedded imposes signaling constraints on the correlations between the experiments. However, it was shown that when the local experiments are described by quantum mechanics, their correlations could violate causality, as defined by the signaling constraints. The latter can be thought as causal inequalities which can be violated by certain scenarios. It was later showed that such correlations can be obtained in a multipartite setting even if the local operations are classical [10], something impossible in the bipartite case [2].

Another example in which causal order is a not well-defined, is the quantum switch [5]. It is an attempt to make a quantum circuit such that the causal order between two events becomes 'indefinite' in the quantum sense. More precisely, local quantum operations are applied in an order that depends on the value of a variable prepared in a quantum superposition [5–7, 11, 18]. This approach allows one to achieve certain tasks that are impossible if the quantum operations are applied in a definite causal order. In contrast to the violation of a causal inequality, however, this conclusion depends on the assumed description of the local operations and is therefore theory-dependent.

The process matrix framework and the quantum switch are examples in which a theory or an implementation seem to be in conflict with causality. However, the analysis of these effects has relied on semi-rigorous considerations about causality or what it means for a scenario to be compatible with 'definite causal order'. A fully rigorous examination of these effects requires a clear notion of causality, which so far, in a background-independent setting, has been lacking. To make this notion general and applicable to any number of parties, turned out to be nontrivial. The difficulty will appear shortly. From simple considerations of the multipartite case, we understand that the causal order of a set of local experiments can be not only random but also depend on the settings of some experiments. This is because the setting of a given experiment can influence the order in which future experiments occur. This motivates a notion of causality expressed as a rule that provides constraints on the joint probabilities for the events in the local experiments but also allowing for the possibility that different future causal configurations unfold depending on past events. A theory of such dynamical causal order is essential not only for the understanding the subject of indefinite causal order in quantum mechanics or more general theories, but also for the problem of inferring causal structure beyond the classic paradigm of underlying deterministic variables and static causal relations [28].

Synopsis of the results: In this chapter, we develop rigorous theory-independent and theory-dependent notions of causality in the process framework and characterize the structure and relations between the corresponding classes of processes they define. The next section (3) is devoted to the theory-independent perspective, which contains our core result. We formalize the process framework in theory-independent terms and propose a definition of causality which allows for the possibility of dynamical causal order. We develop a number of concepts, such as multipartite signaling, reduced and conditional processes, and derive necessary and sufficient conditions for a process to be causal, which are expressed in the form of an iteratively defined canonical decomposition of the probabilities in the process. This decomposition can be understood as describing a causal 'unraveling' of the events in the experiment in a sequence, showing that the proposed notion of causality yields the structure expected from intuition. Apart from being logically non-trivial, this result has important conceptual implications—it presents us with an understanding of causal order as a random function on random events rather than the ordering of underlying locations in which events happen. This perspective is in the spirit of the idea of background independence in general relativity, according to which there are no underlying locations, but only events and the relations between them. In Sect. 2.4, we focus on the quantum process framework, where we develop different theory-dependent notions of causality, which in principle have analogues in more general process theories too. Specifically, we investigate several possible generalizations of the bipartite notion of causal separability, which was previously defined heuristically in the bipartite case by postulating a particular form of the quantum process matrix [2]. We show that this form can be understood as arising from the canonical decomposition of causal processes under the condition that each process in this decomposition is a valid quantum process. We define the multipartite concept based on this principle. We show that the sets of causal and causally separable processes are not equivalent in the multipartite case, by giving an explicit example of a class of processes that are causal but not causally separable. This example is based on the 'quantum switch' technique discussed earlier. We also show that, surprisingly, there exist causally separable (and hence causal) quantum processes that become non-causal if extended by supplying the parties with an entangled input ancilla. This example of 'activation of non-causality' is constructed based on a suitable modification of the non-causal process matrix of Ref. [2]. This observation motivates the concepts of *extensibly causal* and *extensibly causally separable* (ECS) processes, for which the respective property remains invariant under extension with arbitrary input ancillas. We derive a characterization of the class of ECS quantum processes in the tripartite case in terms of simple conditions on the form of the process matrix, which generalize the known form of bipartite causally separable process matrices. In the bipartite case, causal separability and extensible causal separability are equivalent, hence the class of ECS processes can be regarded as another possible multipartite generalization of the previously known bipartite concept. Finally, we consider the class of processes realizable by classically controlled quantum circuits, which we show is inside the class of ECS processes. These, too, are equivalent to the causally separable processes in the bipartite case and provide a possible multipartite generalization based on a different principle. We conjecture that the processes that can be obtained by classically controlled quantum circuits are equivalent to the ECS processes, and hence are described by process matrices obeying the simple conditions we have derived. We provide arguments in favor of this conjecture based on analysis in the tripartite case. In Sect. 2.5, we summarize our results and discuss future research directions.

2.3 The Process Framework

2.3.1 General Processes

Synopsis: In this section we lay down the main concepts of the process framework, which are the same as when formulating a general probabilistic theory: setting, outcome, event, operation, local experiment. The main scene is the following: we want to describe a situation where a number of experiments take place, in some unknown circumstances regarding their causal order. To formally describe this situation, we define a *process*: the collection of probabilities of the local outcomes of the experiments given their settings. This process will be our central tool to explore causality: we will see how causality manifests itself through the constraints that imposes on the process.

The process framework introduced in Ref. [2] describes probabilities of outcomes of local experiments associated with different parties, without assuming a global causal order between the experiments, but only locally, for the events within each experiment. The framework was developed for the case where the local experiments are described by quantum mechanics, under two specific assumptions: the joint probabilities of the outcomes of the local experiments are non-contextual functions of the operations of the parties, and that the local experiments can be extended to act on ancillas prepared in any joint quantum state.

However, it is not necessary to restrict to any particular local theory that describes the local experiments. We formulate a general process framework in operational terms, without specifying the local theory, and therefore without any other additional assumptions that would restrict the parties' correlations. The motivation is to focus our attention purely at the effects of causality on correlations between the experiments. Another motivation comes from the concept of causal inequality, which formulates the bound of correlations compatible with an underlying causal structure in theory-independent terms.

We start with some definitions. Each local experiment, say A is described by two variables: a *setting* $s^A \in S^A$, and, for each such setting, A obtains an *outcome* $o_s^A \in O_s^A$ for that setting. The sets O_s^A for each setting s^A can be thought to be identical as we can always extend them with fictitious outcomes that never occur, i.e. $O_s^A \equiv O^A$. An *operation* is a collection of possible events $\{(s^A, o^A)\}_{o^A \in O^A}$ for a fixed value of the setting $s^A \in S^A$ (Fig. 2.1). The occurrence of a set of a experiments $\{A, B, \cdots\}$ and the circumstances in which they take place are conditioned on some variable $w^{A,B,C,\cdots} \in \Omega^{A,B,C,\cdots}$ which we will discuss later.

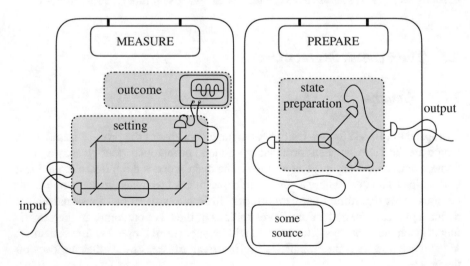

Fig. 2.1 **Local experiment**: a typical local experiment would involve a measurement stage on the input system and a preparation stage of the output system. An operation is the collection of the events defined for every setting and the collection of possible outcomes for that setting

Definition 2.3.1 (*Process*) Mathematically, we define a process $\mathcal{W}^{A,B,\cdots}$ for a set of local experiments (or parties) $\mathcal{S} = \{A, B, \cdots\}$ as the collection of conditional probabilities

$$\mathcal{W}^{A,B,\cdots} \equiv \{P(o^A, o^B, \ldots | s^A, s^B, \ldots, w^{A,B,\cdots})\}, \qquad (2.1)$$
$$o^X \in O^X, s^X \in S^X, X \in \mathcal{S},$$

for a given value of $w^{A,B,\cdots} \in \Omega^{A,B,\cdots}$.

Definition 2.3.2 (*Trivial process*) For the purposes of expressing more succinctly certain conditions later, it is convenient to allow the set of local experiments $\mathcal{S} = \{A, B, \cdots\}$ in the definition to be the empty set $\{\}$ as a special case. In that case, the corresponding process will be referred to as the *trivial process*. We define it to consist of a single probability—that for the trivial outcome given the trivial setting—which is equal to 1.

Note that while in this chapter the word 'process' has a specific meaning, in other chapters it will be used to refer to a general physical process.

A *theory* in the process framework is specified by listing the different types of input and output systems, all possible settings and outcomes, all possible variables $w^{A,B,\cdots}$ for which we have an occurrence of the experiments $\{A, B, \cdots\}$ and the corresponding process (2.1). The variables $s^X, o^X, w^{\{A,B,\cdots\}}$ are identified with their equivalence classes with regard to the probabilities (2.1)—meaning that give rise to the same probabilities. This is also true for the operational probabilistic theories in the circuit framework [29–32].

But what are these variables supposed to describe in practice? In Refs. [16, 17], it was argued that there are two main ideas that underlie the concept of operation in the standard circuit framework for operational probabilistic theories [29–32]. The first one, termed as *closed-box assumption*, is the idea that the input and output systems of an operation are the only means of information exchange responsible for the correlations between the outcomes of that operation and the outcomes of other operations in the global experiment. The second idea, termed as *no-post-selection criterion*, which makes sense assuming a predefined notion of temporal ordering as in the standard circuit formulation, is that the variable that defines an operation, or the setting s^X, can be known with certainty before the time of interaction with the input system unconditionally on any events in the future.

Since no predefined global time is assumed in our picture, the latter condition will be imagined to hold only with respect to the local temporal sequence of events observed by each experimenter. Furthermore, we will assume that the variable $w^{A,B,\cdots}$ that defines the global setup in which the individual experiments take place is also obtained without post-selection. We can make sense of this idea by imagining that the variable is associated with an event that fits within each of the local temporal frames of the experimenters and is such that it occurs before any of them receives the input system. We will call processes that describe experiments of this kind *pre-selected processes*. For a generalization that admits post-selection, see Ref. [16].

For the rest of the chapter, we will consider only pre-selected processes and will refer to them as processes, unless otherwise stated. We will also drop the explicit specification of the variable $w^{\{A,B,\cdots\}}$ on which the global experiment is conditioned. Therefore, we simply write $\mathcal{W}^{A,B,\cdots} \equiv \{p(o^A, o^B, ...|s^A, s^B, ...)\}$ keeping in mind that every process describes circumstances defined by such a variable and hence all probabilities we consider are implicitly conditional on such a variable.

2.3.2   Causal Processes

Synopsis: Causality is well defined in the circuit framework through the strict partial order (SPO) that it defines. In our picture, for a set of local experiments, we have no circuit notion, hence nor SPO, to help us define causality. Nonetheless we can ask: is a given process compatible with the existence of a SPO with respect to which causality is satisfied? We formulate this precisely. We come across two difficulties: first, that the SPO may be random—probabilistic mixtures of SPOs have to be taken into consideration; second, the setting of an experiment could be correlated with the SPO of this experiment and future ones. Therefore, causality must be compatible with a probabilistic SPO and impose restrictions on the settings and the SPO dependence. Considering these, we provide a definition of a process that respects causality: a causal process.

In the circuit framework for operational probabilistic theories, causality is defined as the property that the probability distribution over the outcomes of a given operation in a circuit does not depend on what operations take place in the absolute future or absolute elsewhere [36] of that operation as defined by the *strict partial order* (SPO) of the circuit composition [30, 31]. More specifically, every circuit describes a set of operations taking place at the vertices of a Directed Acyclic Graph (DAG), whose directed edges (the circuit 'wires') correspond to systems that go from one operation to another. Such a graph defines a SPO on the operations in a circuit (a precise definition of SPO is given below)—one operation is in the *absolute past* of another (equivalently, the latter is in the *absolute future* of the former) if there exists a directed path from the former to the latter through the graph. If there is no directed path connecting two operations, we say that one is the *absolute elsewhere* of the other.

In our picture, where we have a set of local experiments, we can associate each experiment with a vertex of such a DAG. Then the property of causality says that the probabilities for the outcomes of local experiments that are in the causal past or causal elsewhere of a given local experiment cannot depend on the setting of that experiment. A circuit theory that obeys this condition, such as standard quantum theory, is called causal, and for such a theory the SPO defined by the circuit composition can be interpreted as causal order [30, 31]. This interpretation corresponds to the intuitive idea that, if the setting of a local experiment is regarded as up to the 'free choice' of an experimenter, then any correlations between that setting and other variables must indicate a causal influence of the setting on those variables. From this perspective, causality can be understood as the condition that a variable can influence only variables in its immediate location or in its absolute future.

In the process framework, we do not assume the existence of a given circuit in which the local experiments are embedded. Thus, there is no natural SPO with respect to which to define causality. Nevertheless, we may ask whether the probabilities described by a given process are compatible with the existence of a SPO with respect to which causality is satisfied. How to formulate this precisely is the main result of this section. However, it is not immediately clear because the process framework can describe situations in which the SPO may be random. For instance, it can describe the correlations between local experiments that can be embedded in different circuits according to some probability distribution. Clearly, if the SPO between the local experiments is random, it must be the case that conditionally on that SPO taking any particular value, the probabilities of the outcomes of the parties given their settings must obey the above notion of causality. This condition, however, is not sufficient to capture the idea of causality.

For example, consider the local experiments of two parties, Alice and Bob, which are embedded at random in one of two possible causal circuits where they occur in different orders. The probabilities for all events and the specific circuit could be such that, conditionally on any particular circuit being realized, the joint probabilities of the outcomes of the parties given their settings obey the above notion of causality, but nevertheless the setting of Alice could be correlated with the circuit in which her experiment is embedded, and thereby with the SPO on the two local experiments. Intuitively, such a situation should be in conflict with causality, because if Alice's setting could not influence events that occur in the past, it should not influence whether or not Bob performs an operation in the past. (Note that, in principle, Alice could change whether or not Bob's operation is in her past simply by waiting. However we do not allow this because we assume that the parties have no access to any global time reference frame, except through the systems they receive in their labs.) In the circuit notion of causality, the past is defined assuming a fixed circuit. In the process framework, where no fixed circuit is assumed, we cannot use such a notion of causality to define that Alice's settings should be independent from past events—there is no circuit with which to define past.

This indicates that we need a more general notion of causality that imposes constraints on how the SPO of the local experiments can depend on the parties' settings. A simple possibility is to require that the SPO on the local experiments must be independent of the parties' setting. This condition, however, is too restrictive, because, compatibly with the idea of causality, we can conceive of scenarios where the setting of a given party influences the order in which other parties perform their experiments in that party's absolute future. Thus, a more sophisticated definition of causality is needed for the process framework. We next develop such a definition.

First, let us review the properties of SPO and introduce some terminology. A SPO on a nonempty set of local elements $S = \{A, B, C, \cdots\}$ is a binary relation $\prec$ which satisfies the following conditions: *(1) irreflexivity*—not $A \prec A$; *(2) transitivity*—if $A \prec B$ and $B \prec C$, then $A \prec C$; *(3) anti-symmetry*—if $A \prec B$, then not $B \prec A$. When two local experiments A and B satisfy $A \prec B$ (equivalently, $B \succ A$), we will say that A is in the *absolute past* of B, or that B is in the *absolute future* of A [36]. It will be convenient to introduce the notation $A \not\prec B$ (equivalently, $B \not\succ A$), which means $A \neq B$ and not $A \prec B$, that is, A and B are different and A is not in the absolute past of B (equivalently, B is not in the absolute future of A). We will also introduce the notation $A \not\prec\not\succ B$, which means $A \not\prec B$ and $A \not\succ B$, that is, A and B are different and A is neither in the absolute past nor in the absolute future of B (same holds for B with respect to A). In the case when $A \not\prec\not\succ B$, we will say that A and B are *absolutely independent*, or that A is in the *absolute elsewhere* [36] of B (and similarly, B is in the absolute elsewhere of A).

The SPO on a set $S = \{A, B, \cdots\}$ is described by the list of respective relations for each such pair, which we will denote by $\kappa(A, B, \cdots)$. As discussed above, the SPO $\kappa(A, B, \cdots)$ in terms of which causality would be defined can most generally be random and correlated with the events in these experiments. The notion of causality would impose constraints on the possible correlations. We want these constraints to formalize the following intuition about causality:

The choice of setting in a local experiment cannot affect the occurrence of events in the absolute past or absolute elsewhere of that experiment, nor the SPO on such events and the experiment in question.

We can think that the variables that describe the events in the absolute past and absolute elsewhere of that experiment, and the SPO of these events and the experiment in question, have already been taken particular values *before* the occurrence of that experiment.

Since a process is defined as $\mathcal{W}^{A,B,\cdots} \equiv \{P(o^A, o^B, ... | s^A, s^B, ..., w^{A,B,\cdots})\}$ and does not assume the existence of probabilities for the settings, we will formulate the above constraint at the level of probabilities conditional on the settings. We define this as follows.

Definition 2.3.3 (*Causal process*) A process $\mathcal{W}^{A,B,\cdots} \equiv \{p(o^A, o^B, \cdots | s^A, s^B, \cdots)\}$ for a nonempty set of local experiments $S = \{A, B, \cdots\}$ is called *causal* if and only if there exists a probability distribution

$$p(\kappa(A, B, \cdots), o^A, o^B, \cdots | s^A, s^B, \cdots),$$

$$\sum_{\kappa(A,B,\cdots)} p(\kappa(A, B, \cdots), o^A, o^B, \cdots | s^A, s^B, \cdots) = p(o^A, o^B, \cdots | s^A, s^B, \cdots),$$

$$(2.2)$$

where the random variable $\kappa(A, B, \cdots)$ takes values in the possible SPOs on $S = \{A, B, \cdots\}$, such that for every local experiment, say A, every subset $\mathcal{X} = \{X, Y, \cdots\}$ of the rest of the local experiments, and every SPO $\kappa(A, X, Y, \cdots) \equiv \kappa(A, \mathcal{X})$ on the local experiment in question and that subset, we have

$$p(\kappa(A, \mathcal{X}), A \not\prec \mathcal{X}, o^{\mathcal{X}} | s^A, s^B, \cdots) = p(\kappa(A, \mathcal{X}), A \not\prec \mathcal{X}, o^{\mathcal{X}} | s^B, \cdots). \quad (2.3)$$

Here $o^{\mathcal{X}}$ denotes collectively the outcomes of all local experiments in $\mathcal{X}$, and $A \not\preceq \mathcal{X}$ denotes the condition that all these local experiments are in the causal past or causal elsewhere of A.

Note. A monopartite process is trivially causal.

For a process $\mathcal{W}^{A,B,\cdots}$ that is causal, the binary relation $\prec$ of the SPO $\kappa(A, B, \cdots)$ can be interpreted as *causal order*. In that case, we will use the terms '*causal past*', '*causal future*', '*causal elsewhere*' and '*causally independent*' in the place of 'absolute past', 'absolute future', 'absolute elsewhere' and 'absolutely independent', respectively. We will also refer to the list of pairwise relations $\kappa(A, B, \cdots)$ as the *causal configuration* of the local experiments.

Our goal next is to understand the structure of causal processes that arises from this definition and show that it corresponds exactly to what one expects from intuition.

2.3.3 Fixed-Order Causal Processes, (No) Signaling, Reduced and Conditional Processes

Synopsis: In this section we provide several definitions. A *fixed-order causal process* is a causal process that is compatible with a single SPO between the parties. We define *bipartite* and *multipartite signaling* and we introduce two types of processes: considering two complementary subsets of parties $\mathcal{A}$ and $\mathcal{B}$, with $\mathcal{B} \not\preceq \mathcal{A}$, a *reduced process* for $\mathcal{A}$ is a process defined only for that subset, independent of what happens in $\mathcal{B}$, denoted as $\mathcal{W}^{\mathcal{A}}$; a *conditional process* is a collection of processes for the set $\mathcal{B}$ conditioned on each event in $\mathcal{A}$, denoted as $\mathcal{W}^{\mathcal{B}|\mathcal{A}}$. The relation between the whole process for the unification of the sets and the reduced and conditional processes is then written as: $\mathcal{W}^{\mathcal{A},\mathcal{B}} \equiv \mathcal{W}^{\mathcal{B}|\mathcal{A}} \circ \mathcal{W}^{\mathcal{A}}$, where the product $\circ$ denotes multiplication of the respective probabilities of these processes, when defined, for the same value of the event in $\mathcal{A}$ (Fig. 2.2). Before we consider the case of general causal processes, it will be instructive to investigate the special case of causal processes for which the causal configuration of the local experiments is fixed. As we will show, the constraints on such processes can be expressed via the concept of signaling, which we develop below. We also introduce several related concepts that will be of use later.

Definition 2.3.4 (*Fixed-order causal process*) A process $\mathcal{W}^{A,B,\cdots} \equiv \{p(o^A, o^B, \cdots \mid s^A, s^B, \cdots)\}$ is called *fixed-order causal* if it is *compatible* with a deterministic causal configuration, i.e., if it satisfies condition (2.3) for a SPO $\kappa(A, B, \cdots)$ that takes a particular value $\kappa(A, B, \cdots) = \kappa_*(A, B, \cdots)$ with unit probability for all possible settings of the parties:

$$p(\kappa(A, B, \cdots), o^A, o^B, \cdots \mid s^A, s^B, \cdots) = 0,$$
$$\text{iff} \quad \kappa(A, B, \cdots) \neq \kappa_*(A, B, \cdots),$$
$$\forall s^A \in S^A, \forall s^B \in S^B, \cdots, \forall o^A \in O^A, \forall o^B \in O^B, \cdots. \tag{2.4}$$

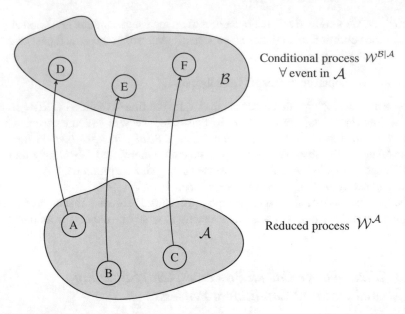

Fig. 2.2 Reduced and conditional process: the arrows represent an example of possible singling between the parties, to indicate that $\mathcal{B} \not\prec \mathcal{A}$

Definition 2.3.5 (*Bipartite signaling*) We say that there is *no signaling* from A to B in a bipartite process $\mathcal{W}^{A,B}$ if and only if the probabilities of the process satisfy

$$p(o^B|s^B, s^A) \equiv \sum_{o^A \in O^A} p(o^A, o^B|s^B, s^A) = p(o^B|s^B), \qquad (2.5)$$

$$\forall s^A \in S^A, s^B \in S^B, o^B \in O^B,$$

i.e., the marginal probabilities for the outcomes of B are independent of the setting of A for any possible setting of B. Equivalently, we say that there *is* signaling from A to B if and only if this condition is not satisfied.

For a fixed-order causal process $\mathcal{W}^{A,B}$, where one of the relations $A \prec B$, $B \prec A$, or $A \not\prec\not\succ B$ holds with unit probability for all settings of the parties, we can see that signaling is possible from one experiment to the other only if the former is in the causal past of the latter, which agrees with the notion of causality in the circuit framework [30, 31].

In the case of more than two local experiments, the relevant generalization of the above notion of signaling may not be immediately obvious. Notice that if a given bipartite process $\mathcal{W}^{A,B}$ involves no signaling between A and B, such a process is in principle compatible with the causal configuration $A \not\prec\not\succ B$ (in fact, it is compatible with any causal configuration of the two parties). However, in the case of processes for more than two local experiments, even if there is lack of signaling between any pair of experiments for every possible settings of the rest of the experiments, the process may not be compatible with a causal configuration in which all experiments are causally independent.

What follows is an example of the subtleties of the multipartite signaling: Consider three local experiments performed by Alice, Bob, and Charlie, where each party's input and output systems are classical bits, and each party is allowed to perform any classical stochastic operation from the input bit to the output bit. Let the experiments of Bob and Charlie be causally independent, and let Alice's experiment be in the absolute future of Bob's experiment, but in the absolute elsewhere of Charlie's experiment (i.e., the causal configuration of the three parties is $[B \prec A, A \not\prec\not\succ C, B \not\prec\not\succ C]$). Imagine that Charlie receives his input system in one of the two possible states 0 or 1 with probability $1/2$, and depending on that state, Alice and Bob are in one of the following two scenarios. In the first scenario (say, when Charlie receives 0), Bob receives a random bit as an input system, his output bit is sent unaltered into the input system of Alice, and Alice's output bit is discarded. In the second scenario (when Charlie receives 1), Bob again receives a random input bit, but this time his output bit is flipped before sending it into Alice's input, and Alice's output bit is again discarded. In both cases, the output system of Charlie is discarded. Clearly, the described situation can be realized in agreement with a fixed causal configuration of the parties—all we need to do is supply Bob with a random bit and correlate the channel from Bob to Alice with the input system of Charlie, discarding the outcomes of Alice and Charlie. The mechanism realizing this is sketched in Fig. 2.3a.

Note that the tripartite process corresponding to this scenario would involve no signaling from Bob to Alice in spite of the existence of a channel from Bob to Alice. This is the case irrespectively of what operation Charlie performs. Obviously, there can be no signaling from Alice to Bob either, since Alice operates in the future of Bob, nor can there be signaling between Alice and Charlie, or between Bob and

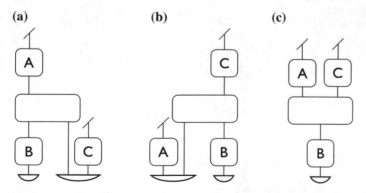

(a)   **(b)**   **(c)**

Fig. 2.3 Subtleties of multipartite signaling: certain types of multipatite signaling correlations do not involve bipartite signaling and do not imply the existence of a causal connection between any particular pairs of channels. The example discussed in the text could arise from any of the mechanisms sketched here [1]

Charlie, since Charlie is causally independent of both Alice and Bob. Thus, we have no signaling between any pair of parties, no matter what the setting of the third party is. Yet, the possible correlations between the parties cannot be realized if all parties are causally independent because if Alice and Charlie measure their input bits and collect the results of their measurements, they can infer the bit sent out by Bob, which is impossible if all parties are causally independent. We might say that in this case we have signaling from Bob to Alice and Charlie together. But intuitively, given the described scenario, this signaling should be from Bob to Alice only, since there is no channel connecting Bob's output system to Charlie's input. However, the latter conclusion is based on knowledge about the mechanism by means of which the correlations are established, or about the causal configuration of the parties, and does not follow solely from the correlations between them. Indeed, the tripartite joint probabilities for the outlined scenario are symmetric with respect to interchanging the roles of Alice and Charlie, and thus they could arise from a different mechanism in a situation where Alice is causally independent of both Bob and Charlie, and Charlie is in the causal future of Bob (Fig. 2.3b). They could also arise from a channel from Bob to both Alice and Charlie (Fig. 2.3c) which transforms Bob's output bit into either correlated or anti-correlated random input bits for Alice and Charlie. We therefore see that, at the level of the joint probabilities for the parties' experiments, there is no way of distinguishing between these different mechanism of information transmission, and hence no way of giving a definition of signaling among a proper subset of the parties that unambiguously captures the existence of such a mechanism. We can, however, give an unambiguous definition of lack of signaling between two complementary subsets of the parties (Fig. 2.4), as well as an associated notion of multipartite signaling, generalizing the bipartite case.

Fig. 2.4 Multipartite signaling: pictorial representation of the definition above

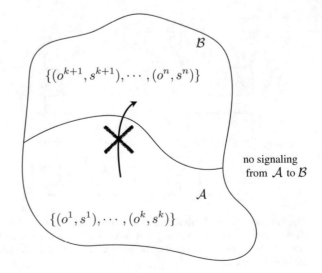

Definition 2.3.6 (*Multipartite signaling*) Consider an n-partite process $\mathcal{W}^{1,\cdots,n}$ for a set of local experiments $\mathcal{S} = \{1, \cdots, n\}$, $n = 0, 1, \cdots$. Let $\mathcal{A} = \{1, \cdots, k\}$ and $\mathcal{B} = \{k+1, \cdots, n\}$, $0 \leq k \leq n$, be two complementary subsets of the experiments, $\mathcal{A} \cup \mathcal{B} = \mathcal{S}$, $\mathcal{A} \cap \mathcal{B} = \{\}$ (for simplicity, we take them to be the first k and the next $n - k$ experiments, which can always be ensured by relabeling). We say that there is *no signaling* from the subset $\mathcal{A}$ to the complementary subset $\mathcal{B}$ in the process $\mathcal{W}^{1,\cdots,n}$ if and only if

$$p(o^{k+1}, \cdots, o^n | s^1, \cdots, s^n) \equiv p(o^{k+1}, \cdots, o^n | s^{k+1}, \cdots, s^n), \qquad (2.6)$$

$$\forall s^j \in S^j, o^j \in O^j, j = 1, \cdots, n.$$

Equivalently, we say that there *is* signaling from (1 or $\cdots$ or k) to ($k+1$ or $\cdots$ or n) if and only if this condition is not satisfied.

Note that this definition only says whether there is signaling from one or more local experiments from a given subset to one or more local experiments from the complementary subset, but in the general case it does not identify pairs of experiments between which there is signaling. In the case of two experiments, the definition reduces to the notion of bipartite signaling defined earlier.

Definition 2.3.7 (*Non-signaling process*) A process $\mathcal{W}^{1,\cdots,n}$ for a set of local experiments $\mathcal{S} = \{1, \cdots, n\}$, $n = 0, 1, \cdots$, is called *non-signaling* if and only if there is no signaling from $\mathcal{A}$ to $\mathcal{B}$ for any pair of complementary subsets $\mathcal{A}$ and $\mathcal{B}$ of $\mathcal{S}$.

From the definition of causal process, one easily obtains the following relation between the existence of multipartite signaling among the local experiments described by a given process and the causal configuration of these experiments.

Proposition 2.3.1 *In an n-partite fixed-order process $\mathcal{W}^{1,\cdots,n}$, $n \geq 1$, compatible with a deterministic causal configuration $\kappa_*(1, \cdots, n)$, there can be signaling from (1 or $\cdots$ or k) to ($k+1$ or $\cdots$ or n), only if at least one of $\{1, \cdots, k\}$ is in the absolute past of at least one of $\{k+1, \cdots, n\}$ according to $\kappa_*(1, \cdots, n)$.*

It turns out that we can formulate necessary and sufficient conditions for a process to be fixed-order causal, which are expressed entirely in terms of the condition stated in Proposition 2.3.1 applied to different subsets of the experiments. To formulate the conditions precisely, we will need to introduce the concept of *reduced process*.

Definition 2.3.8 (*Reduced process*) Consider an n-partite process $\mathcal{W}^{1,\cdots,n}$, $n \geq 0$, for a set of local experiments $\mathcal{S} = \{1, \cdots n\}$. Let $\mathcal{A} = \{1, \cdots, k\}$ and $\mathcal{B} = \{k+1, \cdots, n\}$, $0 \leq k < n$, be two complementary subsets of the experiments (specified up to relabeling), such that there is no signaling from $\mathcal{B}$ to $\mathcal{A}$. This means that

$$p(o^1, \cdots, o^k | s^1, \cdots, s^n) = p(o^1, \cdots, o^k | s^1, \cdots, s^k), \tag{2.7}$$
$$\forall s^j \in S^j, o^j \in O^j, j = 1, \cdots n,$$

The collection of these probabilities will be called *reduced process* for $\mathcal{A}$ and will be denoted by $\mathcal{W}^{\mathcal{A}} \equiv \mathcal{W}^{1, \cdots, k}$.

Note that if a multipartite process is a valid pre-selected process, any of its reduced processes is also a valid pre-selected process because it is defined conditionally on the same pre-selected event. Note also that a general multipartite process need not admit any reduced processes apart from the trivial process and itself, since it may involve signaling from every proper subset of the local experiments to its complementary subset.

Before we state the conditions for a process to be fixed-order causal, we introduce another concept that will be needed later.

Definition 2.3.9 (*Conditional process*) Consider an n-partite process $\mathcal{W}^{1, \cdots, n}$, $n \geq 0$, for a set of local experiments $\mathcal{S} = \{1, \cdots, n\}$. Let $\mathcal{A} = \{1, \cdots, k\}$ and $\mathcal{B} = \{k + 1, \cdots, n\}, 0 \leq k < n$, be two complementary subsets of the experiments (specified up to relabeling), such that there is no signaling from $\mathcal{B}$ to $\mathcal{A}$ (and hence we can define a reduced process $\mathcal{W}^{\mathcal{A}} \equiv \mathcal{W}^{1, \cdots, k}$). For each fixed event $(s^1, o^1, \cdots s^k, o^k)$ in $\mathcal{A}$ for which $p(o^1, \cdots, o^k | s^1, \cdots, s^k) \neq 0$, consider the collection of conditional probabilities $\{p(o^{k+1}, \cdots, o^n | s^{k+1}, \cdots, s^n, s^1, o^1, \cdots, s^k, o^k)\}$. These can be thought of as an $(n - k)$-partite process for $\mathcal{B}$ dependent on the event $(s^1, o^1, \cdots, s^k, o^k)$ in $\mathcal{A}$. The collection of these processes for all values of $(s^1, o^1, \cdots, s^k, o^k)$ for which $p(o^1, \cdots, o^k | s^1, \cdots, s^k) \neq 0$ will be called *conditional process* and will be denoted by $\mathcal{W}^{\mathcal{B}|\mathcal{A}} \equiv \mathcal{W}^{k+1, \cdots, n|1, \cdots, k}$. The relation between the whole process and the reduced and conditional processes can be written in the compact form

$$\mathcal{W}^{\mathcal{A}, \mathcal{B}} \equiv \mathcal{W}^{1, \cdots, n} = \mathcal{W}^{k+1, \cdots, n|1, \cdots, k} \circ \mathcal{W}^{1, \cdots, k}$$
$$\equiv \mathcal{W}^{\mathcal{B}|\mathcal{A}} \circ \mathcal{W}^{\mathcal{A}}, \tag{2.8}$$

where the product $\circ$ between $\mathcal{W}^{\mathcal{B}|\mathcal{A}}$ and $\mathcal{W}^{\mathcal{A}}$ denotes multiplication of the respective probabilities of these processes, when defined, for the same value of the event in $\mathcal{A}$:

$$p(o^1, \cdots, o^n | s^1, \cdots, s^n) = p(o^{k+1}, \cdots, o^n | s^{k+1}, \cdots, s^n, s^1, o^1, \cdots, s^k, o^k)$$
$$p(o^1, \cdots, o^k | s^1, \cdots, s^k), \tag{2.9}$$

for $p(o^1, \cdots, o^k | s^1, \cdots, s^k) \neq 0$, and

$$p(o^1, \cdots, o^n | s^1, \cdots, s^n) = 0, \tag{2.10}$$

for $p(o^1, \cdots, o^k | s^1, \cdots, s^k) = 0$.

Proposition 2.3.2 *A process* $\mathcal{W}^{1,\cdots,n}$ *for a set of local experiments* $\mathcal{S} = \{1, \cdots, n\}$, *$n \geq 1$, is compatible with a deterministic causal configuration* $\kappa_*(1, \cdots, n)$ *of these experiments (and is thereby fixed-order causal) if and only if, for the assumed causal configuration, Proposition 2.3.1 holds for the full process and all of its reduced processes for all bipartitions of the local experiments into two complementary subsets. The proof is given in the Appendix of the related paper [1].*

We next turn to general causal processes, beginning with the bipartite case.

2.3.4 Bipartite Causal Processes

Synopsis: We develop the necessary and sufficient condition for a bipartite process to be causal.

Consider a process $\mathcal{W}^{A,B}$ describing the local experiments of two parties, Alice and Bob. If the process is causal, there exist probabilities $p(A \prec B|s^A, s^B)$, $p(B \prec A|s^A, s^B)$, $p(A \not\prec\not\succ B|s^A, s^B)$, with $p(A \prec B|s^A, s^B) + p(B \prec A|s^A, s^B) + p(A \not\prec\not\succ B|s^A, s^B) = 1$. We can therefore write the joint probabilities of the process in the form

$$
\begin{aligned}
p(o^A, o^B|s^A, s^B) = {} & p(A \prec B|s^A, s^B)\, p(o^A, o^B|s^A, s^B, A \prec B) \\
& + p(B \prec A|s^A, s^B)\, p(o^A, o^B|s^A, s^B, B \prec A) \\
& + p(A \not\prec\not\succ B|s^A, s^B)\, p(o^A, o^B|s^A, s^B, A \not\prec\not\succ B),
\end{aligned}
\tag{2.11}
$$

where each of the probability distributions $p(o^A, o^B|s^A, s^B, A \prec B)$, $p(o^A, o^B|s^A, s^B, B \prec A)$, and $p(o^A, o^B|s^A, s^B, A \not\prec\not\succ B)$, is defined assuming that $p(A \prec B|s^A, s^B) \neq 0$, $p(B \prec A|s^A, s^B) \neq 0$, and $p(A \not\prec\not\succ B|s^A, s^B) \neq 0$, respectively, otherwise that term is absent from the expansion. The definition of causality (2.3) implies that $p(A \prec B|s^A, s^B) = p(A \prec B|s^A)$, $p(B \prec A|s^A, s^B) = p(B \prec A|s^B)$, $p(A \not\prec\not\succ B|s^A, s^B) = p(A \not\prec\not\succ B)$. Since the sum of these probabilities must be unity, we obtain $p(A \prec B|s^A) = p(A \prec B)$, $p(B \prec A|s^B) = p(B \prec A)$, i.e., the causal configuration of the local experiments is independent of the parties' settings. Thus, the probabilities of a bipartite causal process $\mathcal{W}_c^{A,B}$ have the form

$$
\begin{aligned}
p(o^A, o^B|s^A, s^B) = {} & p(A \prec B)\, p(o^A, o^B|s^A, s^B, A \prec B) \\
& + p(B \prec A)\, p(o^A, o^B|s^A, s^B, B \prec A) \\
& + p(A \not\prec\not\succ B)\, p(o^A, o^B|s^A, s^B, A \not\prec\not\succ B),
\end{aligned}
\tag{2.12}
$$

where the probability distributions

$$p(o^A, o^B | s^A, s^B, A \prec B) \equiv p(A \prec B, o^A, o^B | s^A, s^B)/p(A \prec B),$$

$$p(o^A, o^B | s^A, s^B, B \prec A) \equiv p(B \prec A, o^A, o^B | s^A, s^B)/p(B \prec A), \qquad (2.13)$$

$$p(o^A, o^B | s^A, s^B, A \not\prec\not\succ B) \equiv p(A \not\prec\not\succ B, o^A, o^B | s^A, s^B)/p(A \not\prec\not\succ B),$$

whenever defined, describe processes, which we will denote by $\mathcal{W}^{A \prec B}$, $\mathcal{W}^{B \prec A}$, and $\mathcal{W}^{A \not\prec\not\succ B}$, respectively. (Note that we can imagine that the causal configuration $\kappa(A, B)$ taking values $A \prec B$, $B \prec A$, or $A \not\prec\not\succ B$, is associated with an event in the past of both A and B, i.e., the processes $\mathcal{W}^{A \prec B}$, $\mathcal{W}^{B \prec A}$, and $\mathcal{W}^{A \not\prec\not\succ B}$, can be thought of as proper pre-selected processes.) The assumption of causality imposes conditions on these processes too. Specifically, it can be seen that each of them must obey a no-signaling constraint compatible with the concrete causal configuration it is conditioned on: the first one must involve no signaling from Bob to Alice, $p(o^A | s^A, s^B, A \prec B) = p(o^A | s^A, A \prec B)$; the second one must involve no signaling from Alice to Bob, $p(o^B | s^A, s^B, B \prec A) = p(o^B | s^B, B \prec A)$; and the third one must involve no signaling in either direction, $p(o^A | s^A, s^B, A \not\prec\not\succ B) = p(o^A | s^A, A \not\prec\not\succ B)$, $p(o^B | s^A, s^B, A \not\prec\not\succ B) = p(o^B | s^B, A \not\prec\not\succ B)$, i.e., these are fixed-order causal processes. In a compact form, we can write

$$\mathcal{W}_c^{A,B} = p(A \prec B) \, \mathcal{W}^{A \prec B} + p(B \prec A) \, \mathcal{W}^{B \prec A} + p(A \not\prec\not\succ B) \, \mathcal{W}^{A \not\prec\not\succ B}, \qquad (2.14)$$

i.e., a bipartite causal process has the form of a probabilistic mixture of processes that are compatible with the different mutually exclusive causal configurations of the parties (and correspondingly involve only one-way signaling in the respective direction, or no signaling). This form is not only necessary but also sufficient for a process to be causal. This is because the form (2.14) gives explicitly a joint probability distribution $p(\kappa(A, B), o^A, o^B | s^A, s^B) = p(\kappa(A, B)) \, p(o^A, o^B | s^A, s^B, \kappa(A, B))$ that obeys the condition for causality (2.3) when $p(o^A, o^B | s^A, s^B, \kappa(A, B))$ obey the no-signaling constraints compatible with $\kappa(A, B)$ (that is, is independent of the settings of one party in our bipartite case). Indeed, we have

$$
\begin{aligned}
p(A \not\prec B, o^B | s^A, s^B) &= p(B \prec A, o^B | s^A, s^B) + p(A \not\prec\not\succ B, o^B | s^A, s^B) \\
&= p(B \prec A) \, p(o^B | s^A, s^B, B \prec A) + p(A \not\prec\not\succ B) \, p(o^B | s^A, s^B, A \not\prec\not\succ B) \\
&= p(B \prec A) \, p(o^B | s^B, B \prec A) + p(A \not\prec\not\succ B) \, p(o^B | s^B, A \not\prec\not\succ B) = p(A \not\prec B, o^B | s^B),
\end{aligned}
$$

$$(2.15)$$

and similarly $p(B \not\prec A, o^A | s^A, s^B) = p(B \not\prec A, o^A | s^A)$.

Since the non-signaling probabilities $p(o^A, o^B | s^A, s^B, A \not\prec\not\succ B)$ are compatible with the one-way signaling constraints for the cases $A \prec B$ or $B \prec A$, we can also write the probabilities (2.12) in the non-unique form

$$p(o^A, o^B | s^A, s^B) = p(w^{A \not\prec B}) \, p(o^A, o^B | s^A, s^B, w^{A \not\prec B}) + p(w^{B \not\prec A}) \, p(o^A, o^B | s^A, s^B, w^{B \not\prec A}), \qquad (2.16)$$

where $w^{A \not\prec B}$ and $w^{B \not\prec A}$ are two mutually exclusive variables for which the experiments of Alice and Bob respect the relations $A \not\prec B$ and $B \not\prec A$, respectively, with the probabilities of these variables satisfying $p(w^{A \nleq B}) + p(w^{B \nleq A}) = 1$. In a compact form, this can be written

$$\mathcal{W}_c^{A,B} = q \, \mathcal{W}^{A \nleq B} + (1-q) \, \mathcal{W}^{B \nleq A}, \quad 0 \le q \le 1, \tag{2.17}$$

where $\mathcal{W}^{Y \nleq X}$ is a process that involves no signaling from Y to X, i.e.,

$$\mathcal{W}^{Y \nleq X} = \mathcal{W}^{Y|X} \circ \mathcal{W}^X. \tag{2.18}$$

The constraint (2.17) (equivalently, (2.16)) provides a means of testing whether a given bipartite process theory is compatible with causal order. For every fixed number of settings and fixed number of outcomes for each party, the joint probabilities satisfying Eq. (2.16) form a convex polytope, which is the convex hull of the polytope of probabilities that involve no signaling from Alice to Bob, and the polytope of probabilities that involve no signaling from Bob to Alice, called the *causal* polytope. The bipartite causal polytope was studied in Ref. [23]. In Chap. 4 we will discuss about the tripartite causal polytope [24].

2.3.5 Form of Tripartite and n-Partite Causal Processes

Synopsis: We formulate a proposition, where we apply the definition of causality to a set of causally ordered parties. As a consequence we make the claim that the probability for a given set of local experiments to be first is independent of the settings of all parties. From this claim we derive the form of a tripartite causal process: It is a mixture of processes compatible with one party being first (three terms), with two parties being first (three terms) and one last process where all parties are first; or causally independent. The intuition behind these terms (processes) is that they represent the only situations for which we can assign independent probability weights—their probability to occur is independent of all parties' settings. For these three families of terms (one, two and three parties are first) we derive a necessary and sufficient form for the (causal) process which have a compact and intuitive form. For example, in the case where C is first, it is $\mathcal{W}^{[C]^I} = \mathcal{W}_c^{A,B|C} \circ \mathcal{W}^C$, where $\mathcal{W}_c^{A,B|C}$ is the conditional causal process for A and B, given the events of C, $\mathcal{W}^C$ is the reduced process for C, and the superscript $[C]^I$ denotes that party C is first. We then move to the multipartite case, where we generalize our results about the tripartite case, and provide a definition for multipartite causal processes. In both cases, we briefly discuss that for a fixed number of settings and outcomes of the parties, the causal processes form a polytope whose facets define causal inequalities.

In the case of more than two parties, causal processes need not have the simple form of probabilistic mixtures of fixed-order causal processes with probability weights that are independent of the parties' settings. Such processes are of course causal, but there are more general cases for three parties. There is the possibility that the setting of

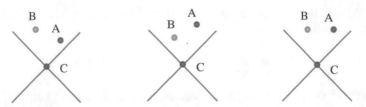

Fig. 2.5 Charlie may change the causal order: in a causal setup where Charlie performs his experiment in the causal past of both Alice and Bob, the causal configuration of Alice and Bob may depend on the setting of Charlie [1]

a party in the past, can affect the causal order of the parties in the future. This is consistent with the notion of causality that we have developed—that the setting of a party cannot affect the causal order of itself and parties in the causal past and causal elsewhere—and consistent with our daily intuition of a *dynamical* causal order.

For a more concrete example, imagine a tripartite experiment, where the parties operate on some internal degree of freedom of a particle that enters their laboratory. Note that the internal degrees of freedom of the particle have to be the only means of exchange of information between the parties. Therefore, the times of input and output of the particle in each laboratory should not be used as an exchange of information. For example, the parties may not be allowed to share any common time reference frame with the rest of the experiment and to perform their operation during a fixed time interval with a stopwatch. In such a case, if Charlie receives a particle first, his operation on the particle could affect the order in which Alice and Bob receive their particles afterwords. For example, conditionally on the outcome of Charlie's experiment, he may send out a particle that is in such a state that takes a different trajectory outside his laboratory than it would have taken had he obtained a different outcome and sent out the particle in a different state. This can result in the different scenarios depicted in Fig. 2.5. By construction, the outlined setup is compatible with the condition that the setting of each local experiment can be chosen independently of events in the causal past and causal elsewhere of that experiment, as well as of the causal configuration of such events and the experiment in question, so it would be associated with a valid causal process.

Clearly, the dependence of the causal configuration of the parties on the parties' settings cannot be arbitrary, because it must agree with causality. To formulate the constraints on this dependence, we will need to introduce some more terminology.

For any fixed causal configuration $\kappa(1, \cdots, n)$ of the local experiments $S = \{1, \cdots, n\}$, there are local experiments that are in no-one else's causal future. The full set of such local experiments, $\{i, j, \cdots\} \subset \{1, \cdots, n\}$, will be referred to as the local experiments that are *first*, or as the *first consecutive set*[1] and will be denoted by

---

[1]In the Chap. 6 following the relevant paper [25] we refer to these sets as non-signaling sets, because by definition the parties in this set cannot signal to each other. However, the consecutive sets are defined by asking the question 'who is first', whereas the nonsignaling sets are defined by the question 'who is last'. This implies that in a given situation, the consecutive sets and the

$[i, j, \cdots]^{\mathrm{I}}$. Next, if the first consecutive set does not include all of the local experiments, there are local experiments whose causal past contains local experiments from, and only from, the first consecutive set. The full set of these will be referred to as the local experiments that are *second*, or as the *second consecutive set*, and will be denoted by $[k, l, \cdots]^{\mathrm{II}}$. Then, if the first and second consecutive sets do not include all local experiments, there are local experiments whose causal past contains local experiments from both sets $[i, j, \cdots]^{\mathrm{I}}$ and $[k, l, \cdots]^{\mathrm{II}}$ and only from those sets. The full set of these will be referred to as the local experiments that are *third*, or as the *third consecutive set*, and will be denoted by $[p, q, \cdots]^{\mathrm{III}}$, and so on.

The following proposition will play a central role in our derivation of the form of multipartite causal processes. We apply the condition for causality (2.3) to consecutive sets. As a reminder, the condition for causality states that for every party A that is not in the causal past of a subset $\mathcal{X}$ of the rest of the parties, the joint probability $p(\kappa(A, \mathcal{X}), A \not\preceq \mathcal{X}, o^{\mathcal{X}} | s^{\mathcal{X}}, s^{A})$ is independent of the settings of A, since A's settings cannot be correlated with any events in its causal past. Now that we have defined the consecutive sets, in the proposition that follows we apply the condition for causality to consecutive sets. The intuition is that, just as the probability of party A to not be in the causal past of $\mathcal{X}$ is independent of the settings of A, the probability for a set of parties to be in one of the first K consecutive sets, is independent of the settings of parties that are in the causal future of this consecutive set (namely that belong to a set with an index larger than K). At the same time, we apply the condition for causality on the parties inside the consecutive set K knowing that all the parties are causally independent, and we get the following proposition.

Proposition 2.3.3 *Consider a causal process for $S = \{1, \cdots, n\}$, $n \geq 1$, with an associated joint probability distribution $p(\kappa(1, \cdots, n), o^1, \cdots, o^n | s^1, \cdots, s^n)$, where $\kappa(1, \cdots, n)$ are the causal configurations of the local experiments. The probability for the first K consecutive sets to consist of specific local experiments, $[1_{\mathrm{I}}, \cdots, n_{\mathrm{I}}]^{\mathrm{I}}, \cdots, [1_K, \cdots, n_K]^K$, these experiments to have a specific causal configuration $\kappa(1_{\mathrm{I}}, \cdots, n_K)$, the experiments in the first $K-\mathrm{I}$ consecutive sets to have a specific set of outcomes $o^{1_{\mathrm{I}}}, \cdots, o^{n_{K-1}}$, and a given (possibly empty) subset $\{1_K, \cdots, g_K\} \subset \{1_K, \cdots, n_K\}$ of the local experiments in the Kth set (given up to relabeling) to have specific outcomes $o^{1_K}, \cdots, o^{g_K}$, can depend non-trivially only on the settings of the local experiments indicated in the first $K-\mathrm{I}$ consecutive sets and the subset $\{1_K, \cdots, g_K\}$,*

$$p(\kappa(1_{\mathrm{I}}, \cdots, n_K), [1_{\mathrm{I}}, \cdots, n_{\mathrm{I}}]^{\mathrm{I}}, \cdots, [1_K, \cdots, n_K]^K, o^{1_{\mathrm{I}}}, \cdots, o^{g_K} | s^1, \cdots, s^n)$$
$$= p(\kappa(1_{\mathrm{I}}, \cdots, n_K), [1_{\mathrm{I}}, \cdots, n_{\mathrm{I}}]^{\mathrm{I}}, \cdots, [1_K, \cdots, n_K]^K, o^{1_{\mathrm{I}}}, \cdots, o^{g_K} | s^{1_{\mathrm{I}}}, \cdots, s^{g_K}),$$
$$(2.19)$$

where we define the 0th set as the empty set. The Proof is given in the Appendix of the related paper [1] and briefly it goes like this: First, observe that the property

---

non-signaling sets will not coincide—although they will both contain sets of parties that are causally independent.

(2.19) holds for the case where the specified K consecutive sets exhaust all local experiments $\{1, \cdots, n\}$. *This is because, in this case, each of the local experiments in the Kth consecutive set is causally preceded by or causally independent from every other local experiment. Hence, the definition of causality (2.3) directly implies the desired relation. The general case follows by induction from this special case and the following Lemma whose proof is given in the Appendix of the related paper [1].*

Lemma 1 *Let the property (2.19) hold for* $K = K' + I$, *where* $K' \geq 1$. *Then it also holds for* $K = K'$.

An important consequence of Proposition 2.3.3 is that the probability for a given set of local experiments to be first is independent of the settings of all parties (this is the case of $K = 1$ and the subset $\{1_K, \cdots, g_K\}$ being empty). For example, consider the different causal configurations of three parties—Alice (A), Bob (B), and Charlie (C)—which are compatible with $[C]^I$ (Fig. 2.5). Each of the individual configurations has a probability that may depend on the setting of Charlie, but the overall probability for Charlie to be first, i.e., for any one of these configurations to be realized (which is the sum of the probabilities for the individual configurations), is independent of the settings of all parties, including Charlie. (To justify the latter, remember that the condition for causality was that the settings of Charlie cannot affect events in its causal past or causal elsewhere nor the causal order of them and Charlie. Hence, if the settings can affect whether a party is first or not, this means that they can affect whether an event occurs in its past or not, which is in conflict with causality). This independence of the first consecutive set on the settings of all parties will play a key role in our characterization of the structure of multipartite causal processes. We will first develop the characterization for the case of three parties in order to illustrate the underlying principle, and then we will extend it to the general multipartite case.

Tripartite Causal Processes

The groups of tripartite causal configurations compatible with the different possibilities for the first consecutive set of parties are listed in Table 2.1. In terms of these possibilities, the probabilities of a tripartite causal process can be written

$$
\begin{aligned}
p(o^A, o^B, o^C | s^A, s^B, s^C) = \; & p([A]^I) \, p(o^A, o^B, o^C | s^A, s^B, s^C, [A]^I) \\
& + p([B]^I) \, p(o^A, o^B, o^C | s^A, s^B, s^C, [B]^I) \\
& + p([C]^I) \, p(o^A, o^B, o^C | s^A, s^B, s^C, [C]^I) \\
& + p([A, B]^I) \, p(o^A, o^B, o^C | s^A, s^B, s^C, [A, B]^I) \\
& + p([A, C]^I) \, p(o^A, o^B, o^C | s^A, s^B, s^C, [A, C]^I) \\
& + p([B, C]^I) \, p(o^A, o^B, o^C | s^A, s^B, s^C, [B, C]^I) \\
& + p([A, B, C]^I) \, p(o^A, o^B, o^C | s^A, s^B, s^C, [A, B, C]^I),
\end{aligned}
\tag{2.20}
$$

Table 2.1 The mutually exclusive groups of tripartite causal configurations

Groups of tripartite causal configurations whose probabilities are independent of the parties' settings, defined by the set of parties that are first
$[A]^{\mathrm{I}}$: $[A \prec B, A \prec C, B \prec C]$ or $[A \prec B, A \prec C, C \prec B]$ or $[A \prec B, A \prec C, B \not\prec\not\succ C]$
$[B]^{\mathrm{I}}$: $[B \prec A, B \prec C, A \prec C]$ or $[B \prec A, B \prec C, C \prec A]$ or $[B \prec A, B \prec C, A \not\prec\not\succ C]$
$[C]^{\mathrm{I}}$: $[C \prec A, C \prec B, A \prec B]$ or $[C \prec A, C \prec B, B \prec A]$ or $[C \prec A, C \prec B, A \not\prec\not\succ B]$
$[A, B]^{\mathrm{I}}$: $[A \not\prec\not\succ B, A \prec C, B \not\prec\not\succ C]$ or $[A \not\prec\not\succ B, A \not\prec\not\succ C, B \prec C]$ or $[A \not\prec\not\succ B, A \prec C, B \prec C]$
$[A, C]^{\mathrm{I}}$: $[A \not\prec\not\succ C, A \prec B, B \not\prec\not\succ C]$ or $[A \not\prec\not\succ C, A \not\prec\not\succ B, C \prec B]$ or $[A \not\prec\not\succ C, A \prec B, C \prec B]$
$[B, C]^{\mathrm{I}}$: $[B \not\prec\not\succ C, B \prec A, C \not\prec\not\succ A]$ or $[B \not\prec\not\succ C, B \not\prec\not\succ A, C \prec A]$ or $[B \not\prec\not\succ C, B \prec A, C \prec A]$
$[A, B, C]^{\mathrm{I}}$: $[A \not\prec\not\succ B, B \not\prec\not\succ C, A \not\prec\not\succ C]$

where

$$
\begin{aligned}
& p([A]^{\mathrm{I}}) + p([B]^{\mathrm{I}}) + p([C]^{\mathrm{I}}) \\
& + p([A, B]^{\mathrm{I}}) + p([A, C]^{\mathrm{I}}) + p([B, C]^{\mathrm{I}}) + p([A, B, C]^{\mathrm{I}}) = 1,
\end{aligned} \tag{2.21}
$$

(see Fig. 2.6 for a pictorial representation of the latter sum) and the probability distributions $p(o^A, ...|s^A, ..., [\cdots]^{\mathrm{I}})$ for a given $[\cdots]^{\mathrm{I}}$, defined whenever $p([\cdots]^{\mathrm{I}}) \neq 0$, describe processes which we will denote by $\mathcal{W}^{[\cdots]^{\mathrm{I}}}$. (Note that we can imagine that the variable $[\cdots]^{\mathrm{I}}$ is associated with an event in the past of all local experiments, i.e., these can be thought of as a proper pre-selected process.)

In a compact form, Eq. (2.20) can be written

$$
\begin{aligned}
\mathcal{W}_c^{A,B,C} = {} & p([A]^{\mathrm{I}})\, \mathcal{W}^{[A]^{\mathrm{I}}} + p([B]^{\mathrm{I}})\, \mathcal{W}^{[B]^{\mathrm{I}}} + p([C]^{\mathrm{I}})\, \mathcal{W}^{[C]^{\mathrm{I}}} \\
& + p([A, B]^{\mathrm{I}})\, \mathcal{W}^{[A,B]^{\mathrm{I}}} + p([A, C]^{\mathrm{I}})\, \mathcal{W}^{[A,C]^{\mathrm{I}}} + p([B, C]^{\mathrm{I}})\, \mathcal{W}^{[B,C]^{\mathrm{I}}} \\
& + p([A, B, C]^{\mathrm{I}})\, \mathcal{W}^{[A,B,C]^{\mathrm{I}}},
\end{aligned} \tag{2.22}
$$

i.e., the overall process is a mixture of processes defined conditionally on the different scenarios $[\cdots]^{\mathrm{I}}$. The processes $\mathcal{W}^{[\cdots]^{\mathrm{I}}}$ cannot be arbitrary but must be compatible with causality, the conditions for which we derive next.

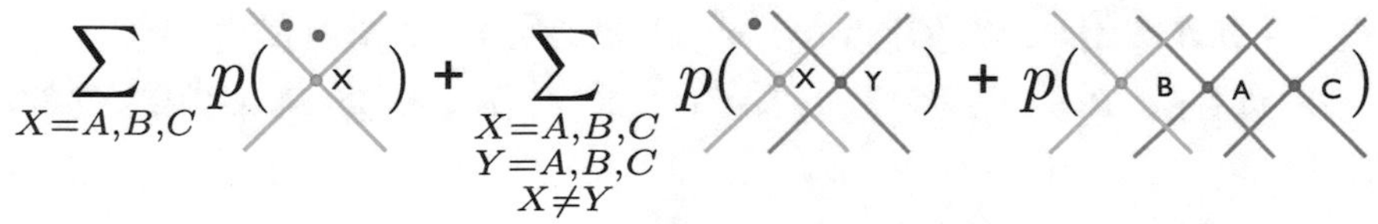
Σ X=A,B,C p(X) + Σ X=A,B,C Y=A,B,C X≠Y p(X Y) + p(B A C)

Fig. 2.6 Pictorial representation of the different independent probabilities of Eq. (2.21)

One party is first: Consider the case in which one party is first, say $[C]^I$ (Fig. 2.5). There are three distinct causal configurations compatible with this case, in which $A \prec B$, $B \prec A$, or $A \npreceq\nsucceq B$ (Table 2.1). We can expand $p(o^A, o^B, o^C | s^A, s^B, s^C, [C]^I)$ conditionally on these configurations as follows:

$$p(o^A, o^B, o^C | s^A, s^B, s^C, [C]^I) = p(o^C | s^A, s^B, s^C, [C]^I) \times$$

$$[p(A \prec B | s^A, s^B, s^C, o^C, [C]^I)\, p(o^A, o^B | s^A, s^B, s^C, o^C, A \prec B, [C]^I) +$$

$$p(B \prec A | s^A, s^B, s^C, o^C, [C]^I)\, p(o^A, o^B | s^A, s^B, s^C, o^C, B \prec A, [C]^I) +$$

$$p(A \npreceq\nsucceq B | s^A, s^B, s^C, o^C, [C]^I)\, p(o^A, o^B | s^A, s^B, s^C, o^C, A \npreceq\nsucceq B, [C]^I)],$$

$$(2.23)$$

where $p(o^A, o^B | s^A, s^B, s^C, o^C, \kappa(A, B), [C]^I)$ is defined when $p(\kappa(A, B) | s^A, s^B, s^C, o^C, [C]^I) \neq 0$, and

$$p(A \prec B | s^A, s^B, s^C, o^C, [C]^I) + p(B \prec A | s^A, s^B, s^C, o^C, [C]^I)$$

$$+ p(A \npreceq\nsucceq B | s^A, s^B, s^C, o^C, [C]^I) = 1. \qquad (2.24)$$

From Proposition 2.3.3, we have that

$$p(o^C | s^A, s^B, s^C, [C]^I) \equiv p([C]^I, o^C | s^A, s^B, s^C) / p([C]^I) = p([C]^I, o^C | s^C) / p([C]^I) = p(o^C | s^C, [C]^I).$$

Similarly, we have

$$p(A \prec B | s^A, s^B, s^C, o^C, [C]^I) = p(A \prec B | s^A, s^C, o^C, [C]^I),$$

$$p(B \prec A | s^A, s^B, s^C, o^C, [C]^I) = p(B \prec A | s^B, s^C, o^C, [C]^I), \qquad (2.25)$$

$$p(A \npreceq\nsucceq B | s^A, s^B, s^C, o^C, [C]^I) = p(A \npreceq\nsucceq B | s^C, o^C, [C]^I),$$

which together with Eq. (2.24) implies

$$p(A \prec B | s^A, s^B, s^C, o^C, [C]^I) = p(A \prec B | s^C, o^C, [C]^I),$$

$$p(B \prec A | s^A, s^B, s^C, o^C, [C]^I) = p(B \prec A | s^C, o^C, [C]^I), \qquad (2.26)$$

$$p(A \npreceq\nsucceq B | s^A, s^B, s^C, o^C, [C]^I) = p(A \npreceq\nsucceq B | s^C, o^C, [C]^I).$$

Substituting this in Eq. (2.23), we obtain

$$p(o^A, o^B, o^C | s^A, s^B, s^C, [C]^I) = p(o^C | s^C, [C]^I) \times$$

$$[p(A \prec B | s^C, o^C, [C]^I)\, p(o^A, o^B | s^A, s^B, s^C, o^C, A \prec B, [C]^I)$$

$$+ p(B \prec A | s^C, o^C, [C]^I)\, p(o^A, o^B | s^A, s^B, s^C, o^C, B \prec A, [C]^I)$$

$$+ p(A \npreceq\nsucceq B | s^C, o^C, [C]^I)\, p(o^A, o^B | s^A, s^B, s^C, o^C, A \npreceq\nsucceq B, [C]^I)], \qquad (2.27)$$

with

$$p(A \prec B|s^C, o^C, [C]^{\mathrm{I}}) + p(B \prec A|s^C, o^C, [C]^{\mathrm{I}}) + p(A \not\prec\not\succ B|s^C, o^C, [C]^{\mathrm{I}}) = 1, \tag{2.28}$$

where the probability distributions

$$p(o^A, o^B|s^A, s^B, s^C, o^C, A \prec B, [C]^{\mathrm{I}}),$$
$$p(o^A, o^B|s^A, s^B, s^C, o^C, B \prec A, [C]^{\mathrm{I}}), \tag{2.29}$$
$$p(o^A, o^B|s^A, s^B, s^C, o^C, A \not\prec\not\succ B, [C]^{\mathrm{I}}),$$

describe bipartite processes for Alice and Bob for every fixed value of (s^C, o^C). The assumption of causality implies conditions for these processes too. They must respect the no-signaling constraints imposed by the causal configuration $\kappa(A, B)$ they are conditioned on—the first one must involve no signaling from Bob to Alice, the second one must involve no signaling from Alice to Bob, and the third one must involve no signaling between Alice and Bob in either direction. This follows from the fact that

$$p(o^A, o^B|s^A, s^B, s^C, o^C, \kappa(A, B), [C]^{\mathrm{I}}) = \frac{p([C]^{\mathrm{I}}, \kappa(A, B), o^A, o^B, o^C|s^A, s^B, s^C)}{p([C]^{\mathrm{I}})\, p(o^C|s^C, [C]^{\mathrm{I}})\, p(\kappa(A, B)|s^C, o^C, [C]^{\mathrm{I}})}, \tag{2.30}$$

and the observation that, on the right-hand side, since only the numerator depends on s^A, o^A, s^B, and o^B, the respective no-signaling constraints on the quantity on the left-hand side follow from the requirement that the numerator is compatible with Eq. (2.3).

To expand our understanding of a tripartite causal process, we use the definitions of reduced and conditional process, to write down a compact and intuitive form for them. Notice that the probabilities $p(o^C|s^C, [C]^{\mathrm{I}})$ in Eq. (2.27) define a reduced monopartite process for Charlie, $\mathcal{W}^C$, while the probabilities enclosed by the square brackets define a conditional bipartite process $\mathcal{W}_c^{A,B|C}$, which is causal (indicated by the subscript c) for every fixed (s^C, o^C). In a compact form, this can be written

$$\mathcal{W}^{[C]^{\mathrm{I}}} = \mathcal{W}_c^{A,B|C} \circ \mathcal{W}^C. \tag{2.31}$$

The form (2.31) is necessary for a causal process for which all causal configurations that have non-zero probabilities respect that $[C]^{\mathrm{I}}$ (in that case, a causal process of the general form (2.22) reduces to the term $\mathcal{W}^{[C]^{\mathrm{I}}}$). It is also sufficient, because this form provides an explicit joint probability distribution $p^{[C]^{\mathrm{I}}}(\kappa(A, B, C), o^A, o^B, o^C| s^A, s^B, s^C)$—equal to

$$p([C]^{\mathrm{I}}, \kappa(A, B), o^A, o^B, o^C|s^A, s^B, s^C) =$$
$$p(o^C|s^C, [C]^{\mathrm{I}})\, p(\kappa(A, B)|s^C, o^C, [C]^{\mathrm{I}})\, p(o^A, o^B|s^A, s^B, s^C, o^C, \kappa(A, B), [C]^{\mathrm{I}}) \tag{2.32}$$

when $\kappa(A, B, C)$ is compatible with $[C]^I$, and to zero otherwise—for which condition (2.3) is satisfied with respect to every party. Indeed, condition (2.3) is satisfied with respect to C since the probability for any party being in the causal past or causal elsewhere of C is zero. It is also satisfied with respect to A (similarly for B) since the no-signaling constraints respected by $p(o^A, o^B | s^A, s^B, s^C, o^C, \kappa(A, B), [C]^I)$ guarantee that

$$p^{[C]^I}(\kappa(A, B, C), A \nprec B, A \nprec C, o^B, o^C | s^A, s^B, s^C) = p^{[C]^I}(\kappa(A, B, C), A \nprec B, A \nprec C, o^B, o^C | s^B, s^C).$$
(2.33)

The necessary and sufficient conditions for a causal process compatible with $[A]^I$ and $[B]^I$ are analogous.

Two parties are first: Let us now consider the case where two parties are first, say $[B, C]^I$. The possible causal configurations in this case (Table 2.1) are depicted in Fig. 2.7. Similarly to the previous case, using the assumption of causality, we can expand the probabilities $p(o^A, ... | s^A, ..., [B, C]^I)$ conditionally on the different configurations as follows:

$$p(o^A, o^B, o^C | s^A, s^B, s^C, [B, C]^I) = p(o^B, o^C | s^B, s^C, [B, C]^I) \times$$
$$[p(B \prec A, C \nprec A | s^B, o^B, s^C, o^C, [B, C]^I) p(o^A | s^A, s^B, o^B, s^C, o^C, B \prec A, C \nprec A, [B, C]^I)$$
$$+ p(B \nprec A, C \prec A | s^B, o^B, s^C, o^C, [B, C]^I) p(o^A | s^A, s^B, o^B, s^C, o^C, B \nprec A, C \prec A, [B, C]^I)$$
$$+ p(B \prec A, C \prec A | s^B, o^B, s^C, o^C, [B, C]^I) p(o^A | s^A, s^B, o^B, s^C, o^C, B \prec A, C \prec A, [B, C]^I)], \quad (2.34)$$

with

$$p(B \prec A, C \nprec A | s^B, o^B, s^C, o^C, [B, C]^I) + (B \nprec A, C \prec A | s^B, o^B, s^C, o^C, [B, C]^I) +$$
$$p(B \prec A, C \prec A | s^B, o^B, s^C, o^C, [B, C]^I) = 1, \quad (2.35)$$

where the probabilities $p(o^B, o^C | s^B, s^C, [B, C]^I)$ in Eq. (2.34) define a reduced bipartite process that involves no signaling between B and C, and the probabilities in the square brackets describe a conditional process for A. The fact that there is no signaling between B and C in the first process follows easily from Proposition 2.3.3.

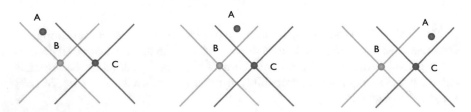

Fig. 2.7 The three possible tripartite causal configurations included in the group where B and C are first. From left to right: $[B \nprec C$ and $B \prec A$ and $C \nprec A]$, $[B \nprec C$ and $B \prec A$ and $C \prec A]$, $[B \nprec C$ and $B \nprec A$ and $C \prec A]$ [1]

It turns out that the decomposition over different causal configurations does not yield any nontrivial conditions on the probabilities of the conditional process enclosed in the square brackets of Eq. 2.34, i.e., the simpler form

$$p(o^A, o^B, o^C | s^A, s^B, s^C, [B, C]^I) = p(o^B, o^C | s^B, s^C, [B, C]^I) \, p(o^A | s^A, s^B, o^B, s^C, o^C, [B, C]^I)$$
(2.36)

is both necessary and sufficient for a valid $\mathcal{W}^{[B,C]^I}$. Necessity is obvious since Eq. (2.34) implies Eq. (2.36). Sufficiency follows from the fact that the right-hand side of Eq. (2.36) is compatible with the particular case $p(B \prec A, C \prec A | s^B, o^B, s^C, o^C, [B, C]^I) = 1$, where the only non-trivial constraints on the probabilities $p(o^A, o^B, o^C | s^A, s^B, s^C, [B, C]^I)$ imposed by $\kappa(A, B, C)$ are that there is no signaling from Alice to Bob and Charlie, and no signaling between Bob and Charlie in their reduced bipartite process. These are clearly guaranteed by Eq. (2.36) when the reduced process $\{p(o^B, o^C | s^B, s^C, [B, C]^I)\}$ involves no signaling between Bob and Charlie. Therefore, similarly to Eq. (2.31), we can write Eq. (2.36) in the compact form

$$\mathcal{W}^{[B,C]^I} = \mathcal{W}^{A|B,C} \circ \mathcal{W}_{ns}^{B,C},$$
(2.37)

where $\mathcal{W}_{ns}^{B,C}$ is a non-signaling bipartite process for Bob and Charlie, and $\mathcal{W}^{A|BC}$ is a monopartite process for Alice conditional on the events in the laboratories of Bob and Charlie.

All three parties are first: Finally, in the case where all of the parties are first, we only have the constraint that

$$\mathcal{W}^{[A,B,C]^I} = \mathcal{W}_{ns}^{A,B,C}$$
(2.38)

is a tripartite non-signaling process. Again, this follows from Proposition 2.3.3.

Result: Form of the tripartite causal processes. We have obtained that a tripartite causal process $\mathcal{W}_c^{A,B,C}$ must have the form

$$\begin{aligned}
\mathcal{W}_c^{A,B,C} = {} & p([A]^I) \, \mathcal{W}_c^{B,C|A} \circ \mathcal{W}^A + p([B]^I) \, \mathcal{W}_c^{A,C|B} \circ \mathcal{W}^B \\
& + p([C]^I) \, \mathcal{W}_c^{A,B|C} \circ \mathcal{W}^C + p([A, B]^I) \, \mathcal{W}^{C|A,B} \circ \mathcal{W}_{ns}^{A,B} \\
& + p([A, C]^I) \, \mathcal{W}^{B|A,C} \circ \mathcal{W}_{ns}^{A,C} + p([B, C]^I) \, \mathcal{W}^{A|B,C} \circ \mathcal{W}_{ns}^{B,C} \\
& + p([A, B, C]^I) \, \mathcal{W}_{ns}^{A,B,C},
\end{aligned}$$
(2.39)

with suitable probability weights $p([A]^I)$, $p([B]^I)$, $p([C]^I)$, $p([A, B]^I)$, $p([A, C]^I)$, $p([B, C]^I)$, and $p([A, B, C]^I)$. This form is also sufficient for a tripartite process to be causal because it explicitly gives a probability distribution

$$p(\kappa(A, B, C), o^A, o^B, o^C | s^A, s^B, s^C) = \sum_{[\cdots]^I} p([\cdots]^I) \, p(\kappa(A, B, C), o^A, o^B, o^C | s^A, s^B, s^C, [\cdots]^I) \quad (2.40)$$

that satisfies Eq. (2.3). Indeed, we have seen that each of the distributions

$$p(\kappa(A, B, C), o^A, o^B, o^C | s^A, s^B, s^C, [\cdots]^I) \tag{2.41}$$

in this convex mixture is an extension of a causal process $\{p(o^A, o^B, o^C | s^A, s^B, s^C, [\cdots]^I)\}$, and hence it satisfies Eq. (2.3). Since the weights $p([\cdots]^I)$ in the mixture are independent of s^A, s^B, and s^C, and Eq. (2.3) is linear in $p(\kappa(A, B, C), o^A, o^B, o^C | s^A, s^B, s^C, [\cdots]^I)$, the equation is satisfied by the whole mixture too.

Condition (2.39) can be further simplified by noticing that the processes corresponding to the cases in which two or three parties are first have forms compatible with cases in which only a single party is first. For instance, $\mathcal{W}^{[B,C]^I}$ satisfies the necessary and sufficient conditions for a valid $\mathcal{W}^{[B]^I}$ or a valid $\mathcal{W}^{[C]^I}$, while $\mathcal{W}^{[A,B,C]^I}$ satisfies the necessary and sufficient conditions for any of $\mathcal{W}^{[A]^I}$, $\mathcal{W}^{[B]^I}$, or $\mathcal{W}^{[C]^I}$. The compatibility of $\mathcal{W}^{[B,C]^I}$ with $[C]^I$, for example, can be seen from the fact that Eq. (2.36) (or Eq. (2.37)) is compatible with the case $[C]^I$ in which $C \prec B \prec A$, since the only constraints in that case are that Alice cannot signal to Bob and Charlie, and that Bob cannot signal to Charlie, which are satisfied by the probabilities in Eq. (2.36). Similarly, $\mathcal{W}^{[B,C]^I}$ is compatible with $[C]^I$. A process $\mathcal{W}^{[A,B,C]^I}$ is compatible with any causal configuration since it does not involve signaling between any of the parties. These observations suggest that we can group (in a generally nonunique way) the terms in the probabilistic mixture (2.22) so as to obtain a mixture of three processes

$$\mathcal{W}_c^{A,B,C} = p(w^{(B,C)\npreceq A}) \, \mathcal{W}^{(B,C)\npreceq A}$$
$$+ \, p(w^{(A,C)\npreceq B}) \, \mathcal{W}^{(A,C)\npreceq B} + p(w^{(A,B)\npreceq C}) \, \mathcal{W}^{(A,B)\npreceq C}, \tag{2.42}$$

where $w^{(B,C)\npreceq A}$, $w^{(A,C)\npreceq B}$, and $w^{(A,B)\npreceq C}$, are some mutually exclusive variables whose probabilities satisfy $p(w^{(B,C)\npreceq A}) + p(w^{(A,C)\npreceq B}) + p(w^{(A,B)\npreceq C}) = 1$, such that conditionally on these variables, the causal configuration of the parties belongs to one of the groups compatible with $(B, C) \npreceq A$ (meaning $B \npreceq A \wedge C \npreceq A$), $(A, C) \npreceq B$, and $(A, B) \npreceq C$, respectively, while the processes $\mathcal{W}^{(B,C)\npreceq A}$, $\mathcal{W}^{(A,C)\npreceq B}$, and $\mathcal{W}^{(A,B)\npreceq C}$, satisfy the most general causal constraints compatible with these groups. For instance, conditionally on $w^{(B,C)\npreceq A}$, the causal configurations of the parties may belong to any of the groups defined by $[A]^I$, $[A, B]^I$, $[A, C]^I$, and $[A, B, C]^I$. The process $\mathcal{W}^{(B,C)\npreceq A}$ would itself be a probabilistic mixture of processes compatible with these groups, which most generally satisfy the constraints satisfied by $\mathcal{W}^{[A]^I}$. That is,

$$\mathcal{W}^{(B,C)\npreceq A} = \mathcal{W}_c^{B,C|A} \circ \mathcal{W}^A, \tag{2.43}$$

$$\mathcal{W}^{(A,C)\npreceq B} = \mathcal{W}_c^{A,C|B} \circ \mathcal{W}^B, \tag{2.44}$$

$$\mathcal{W}^{(A,B)\npreceq C} = \mathcal{W}_c^{A,B|C} \circ \mathcal{W}^C. \tag{2.45}$$

Obviously, the existence of a convex decomposition (2.42) is both necessary and sufficient for a tripartite process to be causal, since any process of the form (2.39) can be written in the form (2.42), while Eq. (2.42) is a special case of Eq. (2.39).

As in the bipartite case, for any fixed number of settings and fixed number of outcomes for each party, the constraint (2.42) provides a means of testing whether the corresponding tripartite probabilities are compatible with causality. The set of probabilities that satisfy Eq. (2.42) is the convex hull of the probabilities compatible with causal configurations in which $(B, C) \not\preceq A$, $(A, C) \not\preceq B$, and $(A, B) \not\preceq C$. As already mentioned, discuss such a tripartite polytope on Chap. 4.

n-Partite Causal Processes

The extension of the conditions for causality of a process to the case of n parties can be defined iteratively. The following theorem provides the generalization of condition (2.39).

Theorem 2 *A process for a set of parties $S = \{1, \cdots, n\}$, $n \geq 1$, is causal if and only if it can be written in the form*

$$W_c^S = \sum_{\mathcal{X} \subset S, \mathcal{X} \neq \{\}} p_{\mathcal{X}} W_c^{S \setminus \mathcal{X} | \mathcal{X}} \circ W_{ns}^{\mathcal{X}}, \qquad (2.46)$$

where the sum is over all nonempty subsets $\mathcal{X}$ of the local experiments S, $p_{\mathcal{X}}$ are suitable probability weights (which can be interpreted as the probability for $\mathcal{X}$ to be first, $p_{\mathcal{X}} = p([\mathcal{X}]^1)$), $S \setminus \mathcal{X}$ denotes the relative complement of $\mathcal{X}$ in S, $W_{ns}^{\mathcal{X}}$ is a non-signaling reduced process for $\mathcal{X}$, and the conditional process $W_c^{S \setminus \mathcal{X} | \mathcal{X}}$ is either the trivial process (when $\mathcal{X} = S$) or otherwise can be written in the same form (2.46) for every given value of the possible events in $\mathcal{X}$. The proof is given in the Appendix of the relevant paper [1].

As in the bipartite and tripartite cases, we can simplify the conditions for an n-partite process to be causal by noticing that the constraints on a process compatible with a given set of k ($1 \leq k \leq n$) parties being first are compatible with the constraints on a process compatible with the case in which only a single one of the k parties is first. Therefore, by an argument analogous to the one in the tripartite case, we obtain the following alternative formulation of the conditions.

Theorem 3 (Canonical causal decomposition) *A causal process for n parties is one that can be written in the (generally non-unique) form*

$$W_c^{1, \cdots, n} = \sum_{i=1}^n q_i W^{(1, \cdots, i-1, i+1, \cdots, n) \not\preceq i}, \quad q_i \geq 0, \forall i, \quad \sum_{i=1}^n q_i = 1, \qquad (2.47)$$

with

$$W^{(1, \cdots, i-1, i+1, \cdots, n) \not\preceq i} = W_c^{1, \cdots, i-1, i+1, \cdots, n | i} \circ W^i, \qquad (2.48)$$

where the $(n-1)$-partite conditional process $W_c^{1,\cdots,i-1,i+1,\cdots,n|i}$ is either trivial (when $n=1$) or has the form (2.47) for every value of the event in i.

The weights q_i in Eq. (2.48) can be thought of as the probabilities $q_i \equiv p(w^{(1,\cdots,i-1,i+1,\cdots,n)\npreceq i})$ for a mutually exclusive set of variables $w^{(1,\cdots,i-1,i+1,\cdots,n)\npreceq i}$ for which the causal configurations of the parties belong to a group such that $(1,\cdots,i-1,i+1,\cdots,n) \npreceq i$.

Theorem 3 (alternatively Theorem 3.4) gives iteratively formulated necessary and sufficient conditions for a process to be causal in the general multipartite case. It can be understood as describing an 'unraveling' of the different possible sequences of operations in steps: first, the party that is first and his/her monopartite process are selected at random based on some probability distribution; next, the party that is second and his/her monopartite process are selected at random from some probability distribution that most generally can depend on the first party's setting and outcome; next, the party that is third and his/her monopartite process are selected from some probability distribution that most generally can depend on the settings and outcomes of the first two parties, and so on. We refer to this intuitive decomposition as the *canonical causal decomposition* of a causal process.

n-partite polytope: By an argument analogous to the one in the tripartite case, one easily sees from Theorem 3 that for any fixed number of settings and outcomes for each party, the causal probabilities for n parties form a polytope, provided that the causal probabilities for $(n-1)$ parties form a polytope. By induction, this implies a polytope structure for the general multipartite case. The nontrivial facets of such a polytope define causal inequalities. Examples of n-partite causal inequalities, where $n = 2k + 1$, for binary inputs and outputs can been found in Refs. [8, 10]. It would be interesting to check if these inequalities are facets of the respective causal polytope.

2.4 The Quantum Process Framework

2.4.1 General Quantum Processes

Synopsis: We've finally reached to the quantum stuff. In this section the process matrix framework is reviewed. We talk about the Hilbert-Schmidt decomposition of the process matrix. This decomposition helps us investigate the process matrix in terms the causal configuration it is compatible with. We write two propositions, one for the allowed terms in a valid process matrix and one on what the terms say about the lack of signaling between sets of parties.

Given the above discussion on our general operational framework for pre-selected processes, or as we call it, the process framework, the quantum process framework, introduced in Ref. [2], is a particular theory within the general process framework. It is based on a set of assumptions about the local operations of the parties and the joint probability of their local outcomes, which we review next.

The first main assumption is that of *local quantum mechanics*, which states that each experimenter inside their laboratory describes their system and operations with quantum mechanics. Specifically, we denote the input and output system of each experimenter by X_I, X_O, with their associated Hilbert spaces $\mathcal{H}^{X_I}$, $\mathcal{H}^{X_O}$ and of dimensions d_{X_I}, d_{X_O} respectively. For each party, the set of (quantum) operations that can be performed describes a quantum instrument [37]. A quantum operation has a set of outcomes $j = 1, \cdots, n$. Each outcome induces a specific transformation from the input to the output system, which is described by a completely positive (CP) map $\mathcal{M}_j^X : \mathcal{L}(\mathcal{H}^{X_I}) \to \mathcal{L}(\mathcal{H}^{X_O})$, where $\mathcal{L}(\mathcal{H})$ is the space of linear operators over the (finite-dimensional) Hilbert space $\mathcal{H}$. The action of each $\mathcal{M}_j^X$ on an operator $\sigma \in \mathcal{L}(\mathcal{H}^{X_I})$ can be written in the Kraus form [38] $\mathcal{M}_j^X(\sigma) = \sum_{k=1}^{m} E_{jk} \sigma E_{jk}^{\dagger}$, $m = d_{X_I} d_{X_O}$, where the Kraus operators $E_{jk} : \mathcal{H}^{X_I} \to \mathcal{H}^{X_O}$ satisfy $\sum_{k=1}^{m} E_{jk}^{\dagger} E_{jk} \le \mathbb{1}^{X_I}$, $\forall j$. The set of CP maps $\left\{ \mathcal{M}_j^X \right\}_{j=1}^{n}$ corresponding to all possible outcomes of a quantum operation has the property that $\sum_{j=1}^{n} \mathcal{M}_j^X$ is CP and trace-preserving (CPTP), which is equivalent to the condition $\sum_{j=1}^{n} \sum_{k=1}^{m} E_{jk}^{\dagger} E_{jk} = \mathbb{1}^{X_I}$.

The second main assumption is that the joint probabilities for the outcomes of the operations of a set of parties, Alice, Bob, Charlie, $\cdots$, is a non-contextual function of the local CP maps,

$$p(i, j, k, \cdots | \{\mathcal{M}_i^A\}, \{\mathcal{M}_j^B\}, \{\mathcal{M}_k^C\} \cdots) = \omega(\mathcal{M}_i^A, \mathcal{M}_j^B, \mathcal{M}_k^C, \cdots). \qquad (2.49)$$

The requirement that local procedures agree with standard quantum mechanics implies that the function ω should be linear in the local CP maps [2].

Such a linear function can be written in a convenient form if we express each local CP map as a positive semidefinite operator using a version of the Choi-Jamiołkowsky (CJ) isomorphism [33, 34]. The CJ operator $M_i^{A_I A_O} \in \mathcal{L}(\mathcal{H}^{A_I} \otimes \mathcal{H}^{A_O})$ corresponding to a linear map $\mathcal{M}_i^A : \mathcal{L}(\mathcal{H}^{A_I}) \to \mathcal{L}(\mathcal{H}^{A_O})$ is defined as $M_i^{A_I A_O} := \left[\mathcal{I} \otimes \mathcal{M}_i \left(|\phi^+\rangle\langle\phi^+| \right) \right]^{\mathrm{T}}$, where $|\phi^+\rangle = \sum_{j=1}^{d_{A_I}} |jj\rangle \in \mathcal{H}^{A_I} \otimes \mathcal{H}^{A_I}$ is a (not normalized) maximally entangled state on two copies of $\mathcal{H}^{A_I}$, the set of states $\{|j\rangle\}_{j=1}^{d_{A_I}}$ is an orthonormal basis of $\mathcal{H}^{A_I}$, $\mathcal{I}$ is the identity map, and T denotes matrix transposition in that basis of A_I and a specific basis of A_O. Using the CJ representation, the joint probabilities (2.49) can be written in the form

$$p(i, j, k, \cdots | \{\mathcal{M}_i^A\}, \{\mathcal{M}_j^B\}, \{\mathcal{M}_k^C\}, \cdots)$$
$$= \mathrm{Tr}\left[W^{A_I A_O B_I B_O C_I C_O \cdots} \left(M_i^{A_I A_O} \otimes M_j^{B_I B_O} \otimes M_k^{C_I C_O} \otimes \cdots \right) \right], \qquad (2.50)$$

The last main assumption behind the quantum process framework is that the local operations of the parties can be extended to act on input ancillas A_I', B_I', C_I', $\cdots$, that are allowed to be prepared in an arbitrary quantum state $\rho^{A_I' B_I' C_I' \cdots}$, where $\rho^{A_I' B_I' C_I' \cdots} \ge 0$, $\mathrm{Tr}\, \rho^{A_I' B_I' C_I' \cdots} = 1$. Upon such an extension, the original operator $W^{A_I A_O B_I B_O C_I C_O \cdots}$ becomes $W^{A_I A_O B_I B_O C_I C_O \cdots} \otimes \rho^{A_I' B_I' C_I' \cdots}$. The requirement that the probabilities are non-negative for any combination of local CP maps $\mathcal{M}_i^A$, $\mathcal{M}_j^B$, $\mathcal{M}_k^C$, $\cdots$, on the extended systems $A = A_I A_I' A_O$, $B = B_I B_I' B_O$, $C = C_I C_I' C_O$, $\cdots$, implies that

$$W^{A_I A_O B_I B_O C_I C_O \cdots} \geq 0. \tag{2.51}$$

Finally, since the probabilities should sum up to 1 for a complete set of local outcomes, we have the following condition

$$\text{Tr}\left[W^{A_I A_O B_I B_O C_I C_O \cdots}\left(M^{A_I A_O} \otimes M^{B_I B_O} \otimes M^{C_I C_O} \otimes \cdots\right)\right] = 1, \tag{2.52}$$

$$\forall M^{A_I A_O}, M^{B_I B_O}, M^{C_I C_O}, \cdots \geq 0,$$

$$\text{Tr}_{A_O} M^{A_I A_O} = \mathbb{1}^{A_I}, \text{Tr}_{B_O} M^{B_I B_O} = \mathbb{1}^{B_I}, \text{Tr}_{C_O} M^{C_I C_O} = \mathbb{1}^{C_I}, \cdots,$$

where Tr_{X_O} denotes partial trace over X_O. Here, we have used the fact that a linear map $\mathcal{M}^X$ is CPTP if and only if its CJ operator satisfies $M^{X_I X_O} \geq 0$ and $\text{Tr}_{X_O} M^{X_I X_O} = \mathbb{1}^{X_I}$. An operator $W^{A_I A_O B_I B_O C_I C_O \cdots}$ that satisfies conditions (2.51) and (2.52) is called a *process matrix* [2]. Knowing the process matrix, by Eq. (2.50) we have the probabilities for the outcomes of any combination of local operations of the parties, i.e., the process matrix provides a complete description of a process. (Here, the set S^X of possible settings of a given party is the set of quantum operations with the respective input and output systems.)

Hilbert-Schmidt decomposition: The process matrix can be expanded in a Hilbert-Schmidt basis of orthogonal matrices on the Hilbert spaces of the input and output systems of the parties, which is helpful in analyzing different properties of the correlations that the process allows. A Hilbert-Schmidt basis of $\mathcal{L}(\mathcal{H}^X)$ is given by a set of Hermitian operators $\{\sigma_\mu^X\}_{\mu=0}^{d_X^2 - 1}$, with $\sigma_0^X = \mathbb{1}^X$, $\text{Tr}\,\sigma_\mu^X \sigma_\nu^X = d_X \delta_{\mu\nu}$, and $\text{Tr}\,\sigma_j^X = 0$ for $j = 1, ..., d_X^2 - 1$. In such a basis, a process matrix can be written

$$W^{A_I A_O B_I B_O C_I C_O \cdots} = \sum_{i,j,k,l,m,n \cdots} w_{ijklmn\cdots}\sigma_i^{A_I} \otimes \sigma_j^{A_O} \otimes \sigma_k^{B_I} \otimes \sigma_l^{B_O} \otimes \sigma_m^{C_I} \otimes \sigma_n^{C_O} \otimes \cdots, \tag{2.53}$$

$$w_{ijklmn\cdots} \in \mathbb{R}, \quad \forall i, j, k, l, m, n, \cdots.$$

It turns out that many properties of process matrices can be formulated entirely as statements about the nonzero terms in the above expansion [2]. For this purpose, it is convenient to introduce the following terminology. Non-zero terms proportional to $\sigma_i^{A_I} \otimes \mathbb{1}^{rest}$ ($i \geq 1$) will be called terms of type A_I, non-zero terms proportional to $\sigma_i^{A_O} \otimes \sigma_j^{B_I} \otimes \mathbb{1}^{rest}$ ($i, j \geq 1$) will be called terms of type $A_O B_I$, etc. Every process matrix also contains a non-zero term proportional to the identity operator on all systems. This term will be referred to as of type $\mathbb{1}$, or as the *identity term*.

Bipartite case—allowed terms: In Ref. [2], it was shown that an operator $W^{A_I A_O B_I B_O}$ satisfies condition (2.52) if and only if it contains at most terms from the following types: $\mathbb{1}$, A_I, B_I, $A_O B_I$, $A_I B_O$, $A_I A_O B_I$, $A_I B_I B_O$. This rule also includes the

Table 2.2 The types of terms that are forbidden in a tripartite process matrix $W^{A_I A_O B_I B_O C_I C_O}$

C_O	$C_I C_O$	B_O	$B_O C_O$
$B_O C_I C_O$	$B_I B_O$	$B_I B_O C_O$	$B_I B_O C_I C_O$
A_O	$A_O C_O$	$A_O C_I C_O$	$A_O B_O$
$A_O B_O C_O$	$A_O B_O C_I C_O$	$A_O B_I B_O$	$A_O B_I B_O C_O$
$A_O B_I B_O C_I C_O$	$A_I A_O$	$A_I A_O C_O$	$A_I A_O C_I C_O$
$A_I A_O B_O$	$A_I A_O B_O C_O$	$A_I A_O B_O C_I C_O$	$A_I A_O B_I B_O$
$A_I A_O B_I B_O C_O$	$A_I A_O B_I B_O C_I C_O$		

Table 2.3 The types of terms allowed in a tripartite process matrix $W^{A_I A_O B_I B_O C_I C_O}$

C_I	$B_O C_I$	B_I	$B_I C_O$	$B_I C_I$
$B_I C_I C_O$	$B_I B_O C_I$	$A_O C_I$	$A_O B_O C_I$	$A_O B_I$
$A_O B_I C_O$	$A_O B_I C_I$	$A_O B_I C_I C_O$	$A_O B_I B_O C_I$	A_I
$A_I C_O$	$A_I C_I$	$A_I C_I C_O$	$A_I B_O$	$A_I B_O C_O$
$A_I B_O C_I$	$A_I B_O C_I C_O$	$A_I B_I$	$A_I B_I C_O$	$A_I B_I C_I$
$A_I B_I C_I C_O$	$A_I B_I B_O$	$A_I B_I B_O C_O$	$A_I B_I B_O C_I$	$A_I B_I B_O C_I C_O$
$A_I A_O C_I$	$A_I A_O B_O C_I$	$A_I A_O B_I$	$A_I A_O B_I C_O$	$A_I A_O B_I C_I$
$A_I A_O B_I C_I C_O$	$A_I A_O B_I B_O C_I$	$\mathbb{1}$		

monopartite case, which is obtained when the input and output systems of one of the parties is trivial (the one-dimensional Hilbert space $\mathbb{C}^1$). Specifically, a monopartite operator $W^{A_I A_O}$ satisfies condition (2.52) if and only if it contains at most terms of type $\mathbb{1}$ and A_I. The types of allowed terms can be generalized to the n-partite case as follows.

Proposition 2.4.1 *An operator of the form* (2.53) *satisfies condition* (2.52) *if and only if in addition to the identity term it contains at most terms in which there is a nontrivial σ operator on X_1 and a trivial one (the identity operator) on X_2 for some party $X \in \{A, B, C, \cdots\}$.*

Tripartite case—allowed terms: In the Appendix of the related paper [1], we present the Proof of the above proposition for the case of three parties and the general case follows accordingly. From the analysis in that Proof we see that a general operator $W^{A_I A_O B_I B_O C_I C_O}$ can contain up to 64 types of terms. The condition for normalisation imposes further constraints. Table 2.2 lists the overall forbidden types of terms, and Table 2.3 lists the allowed types of terms. The positive semidefiniteness condition (2.51) does not limit any further the allowed types of terms, because one can conceive of a positive semidefinite matrix containing nonzero terms of any chosen type (this can be ensured by taking the nontrivial σ terms with non-zero coefficients of sufficiently small magnitude relative to the weight of the identity term which is always fixed). Thus, an operator $W^{A_I A_O B_I B_O C_I C_O}$ is a valid tripartite process matrix,

Table 2.4 The types of terms allowed in a causal process matrix $W^{A_I A_O B_I B_O C_I C_O}_{(A,B)\not\preceq C}$ compatible with $(A, B) \not\preceq C$

C_I	B_I	$B_I C_O$	$B_I C_I$	$B_I C_I C_O$
$A_O B_I$	$A_O B_I C_O$	$A_O B_I C_I C_O$	A_I	$A_I C_O$
$A_I C_I$	$A_I C_I C_O$	$A_I B_O$	$A_I B_O C_O$	$A_I B_O C_I C_O$
$A_I B_I$	$A_I B_I C_O$	$A_I B_I C_I$	$A_I B_I C_I C_O$	$A_I B_I B_O$
$A_I B_I B_O C_O$	$A_I B_I B_O C_I C_O$	$A_I A_O B_I$	$A_I A_O B_I C_O$	$A_I A_O B_I C_I C_O$
$A_I B_O C_I$	$A_O B_I C_I$	$A_I A_O B_I C_I$	$A_I B_I B_O C_I$	$\mathbb{1}$

i.e., it satisfies conditions (2.51) and (2.52), if and only if it satisfies condition (2.51) and contains only terms of the types listed in Table 2.3, where the identity term comes with the weight $w_{000000} = \frac{1}{d_{A_I} d_{B_I} d_{C_I}}$. In a similar way, one proves the allowed types of terms in the general n-partite case. (For an alternative formulation of the conditions for an operator to be a valid process matrix, see Ref. [21].)

n-**partite case - allowed terms**: The types of terms that appear in the expansion of a process matrix are closely related to the signaling between the parties that the process allows. For example, a bipartite process involves signaling from Bob to Alice if and only if the process matrix contains terms of type $A_I B_O$ or $A_I B_I B_O$ [2]. To state the condition for (no) signaling in the multipartite case, it is convenient to introduce the following terminology (see also Ref. [21]). Consider a Hilbert-Schmidt term $\sigma_i^{A_I} \otimes \sigma_j^{A_O} \otimes \sigma_k^{B_I} \otimes \sigma_l^{B_O} \otimes \sigma_m^{C_I} \otimes \sigma_n^{C_O} \otimes \cdots$ as in Eq. (2.53). The *restriction* of this term onto, say, subsystems $B_O C_I C_O \cdots$ is defined as $\sigma_l^{B_O} \otimes \sigma_m^{C_I} \otimes \sigma_n^{C_O} \otimes \cdots$.

Proposition 2.4.2 *An n-partite process matrix for a set of parties $\{1, \cdots, n\}$ does not permit signaling from, say, (1 and 2 and $\cdots$ and k) to ($k + 1$ and $k + 2$ and $\cdots$ and n) if an only if it contains only terms whose restriction onto $1_1 1_2 \cdots k_1 k_2$ are of the allowed types for a process matrix on $\{1, \cdots, k\}$ as described in Proposition 2.4.1. The Proof is given in the Appendix of the related paper [1].*

As an example, a tripartite quantum process that is causal and compatible with a situation in which Charlie is first (Fig. 2.5) should involve no signaling from Alice and Bob to Charlie, and hence it can only contain the types of terms listed in Table 2.4. These constraints on the allowed types of terms imposed by causal order will turn out to play an important role in the characterization of the so-called causally separable quantum processes, which we define in the next subsection.

2.4.2 Causally Separable Quantum Processes

Synopsis: So far we have seen that a quantum process is described by the process matrix. Here, we look at how the property of causality is expressed in terms of simple conditions on the form of the process matrix. When these conditions are satisfied the process is called *causally separable*. We provide the bipartite case as an example and we define the notion of causal separability to the multipartite case.

So far, we have investigated the property of causality in general processes: those that obey it have a particular form and they are called causal processes. For quantum processes, in terms of probabilities, of course nothing changes. A causal quantum process is also causal process, i.e. no matter the specific theory used to describe the local experiments, the collection of probabilities of the local outcomes will be still be a process. Now we can further our investigation of the manifestations of the property of causality: we can look into the process matrix of a quantum causal process and see how the property of causality is expressed in terms of conditions on that matrix.

Bipartite case: Consider a bipartite quantum process for Alice and Bob, and assume that it is a fixed-order process compatible with the causal configuration $A \prec B$. In that case, as argued earlier, the only constraint imposed by causal order is that the process should involve no signaling from Bob to Alice. As pointed out in the previous subsection, there can be signaling from Bob to Alice if and only if the process matrix $W^{A_I A_O B_I B_O}$ contains terms of type $A_I B_O$ or $A_I B_I B_O$. Therefore, a process matrix is compatible with $A \prec B$ if and only if none of these types of terms appear in its expansion. This means that such a process matrix has the form

$$W^{A \prec B} = W^{A_I A_O B_I} \otimes \mathbb{1}^{B_O}, \tag{2.54}$$

where $W^{A_I A_O B_I} \geq 0$ (with $\mathrm{Tr} W^{A_I A_O B_I} = d_{A_O}$) contains at most terms of type $\mathbb{1}$, A_I, B_I, $A_I B_I$, $A_O B_I$, $A_I A_O B_I$. (This is equivalent to saying that $W^{A_I A_O B_I}$ is a valid process matrix for the case where Bob has a trivial output system, $\mathcal{H}^{B_I} = \mathbb{C}^1$.)

Similarly, in the case where $A \not\prec\not\succ B$, the process matrix has the form

$$W^{A \not\prec\not\succ B} = W^{A_I B_I} \otimes \mathbb{1}^{A_O B_O}, \tag{2.55}$$

where $W^{A_I B_I} \geq 0$, $\mathrm{Tr} W^{A_I B_I} = 1$. Such a process is realized in a situation in which Alice and Bob receive input systems in a joint quantum state with a density matrix $W^{A_I B_I}$, and their output systems are discarded.

We can unify these two conditions to write down the form of a process matrix compatible with $B \not\prec A$, which is identical to (2.54),

$$W^{B \not\prec A} = W^{A_I A_O B_I} \otimes \mathbb{1}^{B_O}, \tag{2.56}$$

where $W^{A_I A_O B_I}$ is a valid process matrix for the case where $\mathcal{H}^{B_I} = \mathbb{C}^1$.

As shown in Ref. [39] within a different framework, all process matrices of the type (2.56) can be realized by embedding the experiments of Alice and Bob in a quantum circuit, so that Bob's experiment does not precede Alice's experiment in the order of the circuit composition. Most generally, this corresponds to providing Alice with an input system that is entangled with an ancilla, then sending Alice's output together with the ancilla through a quantum channel into Bob's input, and then discarding Bob's output. Such a process is referred to as quantum 'channel with memory'.

As we have seen earlier, a bipartite causal process is one that can be written in the form (2.17) (which we re-write here for conveniency)

$$W_c^{A,B} = q\, \mathcal{W}^{A \npreceq B} + (1-q)\, \mathcal{W}^{B \npreceq A}, \quad 0 \le q \le 1, \tag{2.57}$$

where $\mathcal{W}^{A \npreceq B}$ and $\mathcal{W}^{B \npreceq A}$ are two processes compatible with $A \npreceq B$ and $B \npreceq A$, respectively. (Remember we denote a process with $\mathcal{W}$ and a process matrix with W). It is then tempting to conjecture that the class of causal quantum processes might be those whose process matrices can be written in the form

$$W^{A_I A_O B_I B_O} = q\, W^{A \npreceq B} + (1-q)\, W^{B \npreceq A}, \quad 0 \le q \le 1, \tag{2.58}$$

where $W^{A \npreceq B}$ and $W^{B \npreceq A}$ have the form defined in Eq. (2.56). Certainly, since the probabilities for the outcomes in the quantum process framework are linear functions of the process matrix, a process matrix of the form (2.58) describes a causal process.

However, the reverse is not necessarily true: a causal process is not necessarily described by a process matrix of the form (2.58). This means that the condition for a process to be causal (Eq. (2.57)) does not imply that $\mathcal{W}^{A \npreceq B}$ and $\mathcal{W}^{B \npreceq A}$ in the convex decomposition of the process should themselves be quantum process; only their convex mixture needs to be. While it is conceivable that the structure of quantum processes might imply the form (2.58) (indeed, this has been shown to hold for a limited class of bipartite quantum processes [15]), there is no obvious reason to expect this to hold in the general case. In fact, we will see that the natural generalization of condition (2.58) to the multipartite case is not equivalent to the condition that a process is causal. Recently, the same was shown to hold also in the bipartite case, by Feix et al. [41].

A bipartite quantum process that admits the decomposition (2.58) was called *causally separable* [2]. One way to think of the relation between causal and causally separable quantum processes is in analogy with the relation between Bell-local and separable (non-entangled) quantum states. A Bell-local state is one for which the joint probabilities for the outcomes of any combination of local measurements admits a local hidden variable description (and hence such a state cannot be used to violate any Bell inequality [42]). A separable quantum state is one for which each of the local distributions can be thought of as arising from the respective local measurements being applied on a local quantum state. A separable quantum state is clearly Bell local, but the reverse is known not to be true [43]. The relation between causal (2.17) and causally separable (2.58) bipartite quantum processes can be seen in an analogous way—a causally separable process is one for which the processes into which we decompose the process are themselves valid quantum processes, which means that the set of causal processes is strictly larger than the one of causally separable processes.

Result: Here, we propose to extend the notion of causal separability to the multipartite case.

Definition 2.4.1 (*Causally separable quantum process*) A quantum process is called causally separable if and only if it can be decomposed in the canonical form given by Theorem 3 (which we re-write below for conveniency), with the additional condition that each process on the right-hand side of Eq. (2.59) is a quantum process. (Note that since the canonical form is defined iteratively, the latter is understood to hold for all conditional processes in this definition).

Theorem (Canonical causal decomposition) *A causal process for n parties is one that can be written in the (generally non-unique) form*

$$\mathcal{W}_c^{1,\cdots,n} = \sum_{i=1}^{n} q_i \mathcal{W}^{(1,\cdots,i-1,i+1,\cdots,n)\not\preceq i}, \quad q_i \geq 0, \forall i, \ \sum_{i=1}^{n} q_i = 1, \tag{2.59}$$

with

$$\mathcal{W}^{(1,\cdots,i-1,i+1,\cdots,n)\not\preceq i} = \mathcal{W}_c^{1,\cdots,i-1,i+1,\cdots,n|i} \circ \mathcal{W}^i, \tag{2.60}$$

where the $(n-1)$-partite conditional process $\mathcal{W}_c^{1,\cdots,i-1,i+1,\cdots,n|i}$ is either trivial (when $n=1$) or has the form (2.59) for every value of the event in i.

By a direct analogy, causally separable processes can be defined for any theory formulated in the process framework, but here we will be interested specifically in quantum processes. The process matrix of a causally separable quantum process will be called a *causally separable process matrix*.

2.4.3 The Quantum Switch: An Example of the Non-equivalence Between Causal and Causally Separable Process in the Tripartite Case

Synopsis: We describe the setup of the quantum switch: a superposition of two causally order quantum circuits, in each of which two parties, A and B, operate one after the other (with alternate orders for each circuit) connected by a channel, and in both cases are succeeded by a third party, C. We prove that the process matrix that describes such a tripartite process is not causally separable, i.e. it cannot be written as a convex combination of process matrices each compatible with $(Y^1, Y^2) \not\preceq Y^3$, for $Y^i = A, B, C$. We show that the quantum process, however, (the collection of the joint probabilities of the local outcomes, given their choice of settings) is causal.

The quantum switch was introduced in [5], as an example of a quantum architecture beyond that of the standard quantum circuit model: a superposition of causally ordered quantum circuits. In that, the order of two black-box quantum operations is conditioned on the value of a qubit (control qubit) prepared in superposition of two logical values. The quantum operations are performed on a target bit, which is the input and output system of the parties and we assume here to be a qubit. We can think that the control bit, which controls the order of the operations, and the target bit on which the operations are performed, are two different degrees of freedom of the same system (like polarization and orbital angular momentum (OAM) respectively of a photon). We can imagine that the particle can go in superposition of two different paths (with the assumed degrees of freedom this can be achieved with a polarizing beam-splitter): one path that goes first through Alice and then though Bob, and one with the reverse order. For simplicity, we can imagine that the particle would always go through Bob at a fixed time and, conditionally on the value of the control bit, the particle would go through Alice before, or after that. It is assumed that independently of the time at which the system may go through Alice' laboratory in a given run, Alice would apply the same operation on it.

To understand the effect of such a setup, consider first the case in which Alice and Bob each apply a unitary operation on the system, U_A and U_B, respectively. Let us denote the Hilbert space of the control qubit (polarization/path degree of freedom) by $\mathcal{H}^c$, and that of the system (OAM degree of freedom) by $\mathcal{H}^s$. Assume that $|0\rangle^c$ corresponds to the path in which Alice is before Bob and $|1\rangle^c$ to the path with the reverse order. If we initially prepare the particle in the state, say, $\rho_{in}^{cs} = |\Psi\rangle\langle\Psi|_{in}^{cs}$, where $|\Psi\rangle_{in}^{cs} = (\alpha|0\rangle^c + \beta|1\rangle^c)|\psi\rangle^s$, at the end it will be in the state $\rho_{fi}^{cs} = |\Psi\rangle\langle\Psi|_{fi}^{cs}$, where $|\Psi\rangle_{fi}^{cs} = \alpha|0\rangle^c U_B^s U_A^s |\psi\rangle^s + \beta|1\rangle^c U_A^s U_B^s |\psi\rangle^s$.

Now, if a third party, Charlie, performs an operation on the joint system $\mathcal{H}^c \otimes \mathcal{H}^s$ subsequently, he can distinguish this situation from a situation in which the order between the operations of Alice and Bob is conditioned on a classical bit (e.g., modeled by the initial state of the control qubit being in a 'classical' mixture of the two possible values, $|\alpha|^2|0\rangle\langle0|^s + |\beta|^2|1\rangle\langle1|^s$, instead of a coherent superposition) by performing a suitable measurement. In fact, it was shown in Ref. [7] that by exploiting such a coherent strategy, Charlie can perfectly distinguish whether a pair of unitaries U^A and U^B commute or anti-commute by using each of the unitaries only once, which is impossible if the order of the unitaries is conditioned on a classical bit. An experimental demonstration of this effect was recently reported in Ref. [18] and in even more recently by the same lab in Ref. [19] where one of the unitaries is replaced by a measure-and-prepare operation. However, both experiments, as they use the same setup for the quantum switch, suffer from some conceptual and experimental loopholes. This will be discussed on the later on, in Chap. 3, where we present an implementation of the quantum switch without such experimental loopholes.

In the general case, the operations of Alice and Bob need not be unitary and may have more possible outcomes. Every such operation, however, can be seen as the result of a joint unitary on the input system and a local ancilla, such that the outcome remains stored on the local ancilla in a particular basis. Similarly, any local 'choice' of operation can be modeled by a larger unitary on all systems involved plus a local ancilla that carries the 'choice' variable. Thus, we can have Alice and Bob perform general operations in this setup by purifying their local operations to unitaries and deferring the reading of their outcomes to the end of the whole experiment. (Note that in order not to destroy the superposition, the whole experiments needs to be performed coherently, which may be unrealistic for local operations performed by macroscopic devices, unless the operations are unitaries, but is in principle compatible with standard quantum mechanics).

In our example, we will take $\alpha = \beta = \frac{1}{\sqrt{2}}$, and we will assume, as described above, that Charlie can operate on both the path and OAM degrees of freedom of the particle after it has interacted with Alice and Bob. In other words, Charlie's input system will be four dimensional, and we will formally decompose it into two qubit subsystems, $\mathcal{H}^{C_I} = \mathcal{H}^{C_I^c} \otimes \mathcal{H}^{C_I^s}$, where $\mathcal{H}^{C_I^c}$ and $\mathcal{H}^{C_I^s}$ correspond to the path and OAM degrees of freedom, respectively. Since Charlie operates last, we do not need to introduce a non-trivial output system for him, i.e., his output system will be assumed one-dimensional. The process matrix relating the local experiment of Alice, Bob, and Charlie in this setup can easily be obtained by describing the experiment in the form of a circuit in which Alice's operation is represented by two controlled operations at two possible times, such that one of them would act nontrivially depending on the state of the control qubit (left diagram on Fig. 2.8). Using the CJ representation of the channels connecting the different boxes, we obtain

$$W^{A_I A_O B_I B_O C_I} = |W\rangle\langle W|^{A_I A_O B_I B_O C_I}, \tag{2.61}$$

where

$$|W\rangle^{A_I A_O B_I B_O C_I} = (|0\rangle^{C_I^c}|\psi\rangle^{A_I}|\Phi^+\rangle^{A_O B_I}|\Phi^+\rangle^{B_O C_I^s} + |1\rangle^{C_I^c}|\psi\rangle^{B_I}|\Phi^+\rangle^{B_O A_I}|\Phi^+\rangle^{A_O C_I^s})/\sqrt{2}, \tag{2.62}$$

with $|\Phi^+\rangle = |00\rangle + |11\rangle$. It can be verified that $W^{A_I A_O B_I B_O C_I C_O}$ contains only allowed terms. The process matrix is a rank-one projector and hence cannot be written as a convex mixture of different process matrices. We now check if it is causally separable: if it is, it must be of one of the types $W^{(A,B) \npreceq C}$, $W^{(B,C) \npreceq A}$, or $W^{(A,C) \npreceq B}$. Each of these matrices is compatible with signaling from the set of two parties to the third one. However, the above process matrix does not fall into any of the three categories: it allows signaling to any of the parties, from some of the other parties. For example, there can be signaling from Alice and Bob to Charlie, as different unitary operations yield different input states for Charlie and therefore different probabilities of the local outcomes for some measurements of Charlie. To see that we can have signaling from Alice to Bob and vice-versa, first notice that Charlie has a trivial output and hence he cannot signal to anyone. This means that we have a well-defined reduced process for Alice and Bob, whose process matrix is

$$W^{A_I A_O B_I B_O} = \frac{1}{2}(|\psi\rangle\langle\psi|^{A_I} \otimes |\Phi^+\rangle\langle\Phi^+|^{A_O B_I} \otimes \mathbb{1}^{B_O} + |\psi\rangle\langle\psi|^{B_I} \otimes |\Phi^+\rangle\langle\Phi^+|^{B_O A_I} \otimes \mathbb{1}^{A_O}). \tag{2.63}$$

This is a causally separable bipartite process matrix describing an equally weighted probabilistic mixture of two fixed-order processes: one where Alice precedes Bob

(Alice receives the input state $|\psi\rangle$, her output is sent to Bob through the identity channel (whose CJ matrix is $|\Phi^+\rangle\langle\Phi^+|^{A_O B_I}$), and his output is discarded; and one process where Bob precedes Alice (with the same mechanism as before with the roles of the parties interchanged). Clearly, since in the first situation there is an ideal channel from Alice to Bob, there can be signaling from Alice to Bob in this process, and similarly from Bob to Alice. Therefore, the process matrix given by Eqs. (2.61) and (2.62) is not causally separable.

However, the quantum process arising from this setup is causal. This follows from the fact that the reduced process for Alice and Bob is causally separable (and hence causal). Specifically, the bipartite process is written as

$$W^{AB} = \frac{1}{2}W^{B \npreceq A} + \frac{1}{2}W^{A \npreceq B} = \frac{1}{2}W^{B|A}_{[A]^I} \circ W^A + \frac{1}{2}W^{A|B}_{[B]^I} \circ W^B \qquad (2.64)$$

and the tripartite process is simply

$$W^{ABC} = W^{C|AB} \circ W^{AB} = \frac{1}{2}W^{C|AB} \circ W^{B|A}_{[A]^I} \circ W^A + \frac{1}{2}W^{C|AB} \circ W^{A|B}_{[B]^I} \circ W^B \qquad (2.65)$$

which is the form of a causal process. This observation suggests how the tripartite joint probabilities of the local outcomes can be simulated without using the quantum switch, if we extend the input and output systems of the parties. In particular, it is clear that the joint probabilities of Alice and Bob can be obtained by a mixture of fixed-order circuits (their process is causally separable) which can be realized by a *classical* switch, where the control bit is classical. To simulate the tripartite joint probabilities, all that is needed is that Charlie receives the information about the local settings and outcomes of the parties so that he makes the right measurements to produce the particular necessary $p(o^C|s^A, o^A, s^B, o^B, s^C)$. Therefore, in addition to the qubit system that goes through Alice and Bob, there has to be another system on which each party writes down their setting and outcome (right diagram on Fig. 2.8), and this system at the end goes to Charlie. Like this, the process can be simulated using a classical mixture of causal configurations.

By a similar argument we can construct a large class of multipartite processes that are causal but not causally separable. Consider a situation in which the order of all but one of the parties is conditioned on the state of a control system prepared in superposition, and subsequently all systems on which these parties have operated together with the control system are sent into the input of the last party. If all systems were initially prepared in a pure state and all channels are unitary ones, the process matrix will have rank 1, and unless the process is fixed-order causal, it cannot be causally separable. Yet, it will be causal because the reduced process for all parties except for the last one will be causally separable (and hence causal) due to the fact that when we trace out the control system, the process for these parties would be a classical probabilistic mixture of fixed-order processes. Since the full process is obtained by multiplying the conditional process of the last party with the reduced process of the previous ones, the full process is causal. It can be simulated using classical control

Fig. 2.8 The left diagram illustrates the circuit with quantum control. The dark circle on the control bit represents a control gate for Alice, and the white circle is also a controlled gate with a bit flip before. The right diagram illustrates a simulation of the same correlations with a classically controlled circuit using input and output systems of larger dimensions, denoted by the extra green system [1]

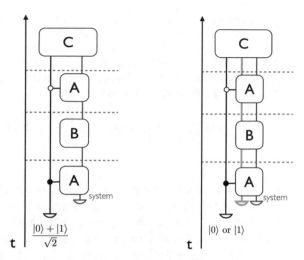

of the order of the parties by allowing larger input and output systems by which the settings and outcomes of all other parties are made available to the last one.

2.4.4 Activation of Non-causality with Shared Entanglement

Synopsis: We present an example of a tripartite causally separable process matrix (giving rise to causal process), whose process becomes non-causal when two parties are supplied with entangled ancillas. This example is surprising and counterintuitive, because the process that corresponds to an entangled input state for two parties is causal—it yields non-signaling correlations.

We show here a rather peculiar property of causality, causal separability and supplied entanglement in the case of quantum processes. Remember that one of the key assumptions in the process matrix framework is that every quantum process can be extended by supplying the parties with ancillas in an arbitrary quantum state, yielding another valid process. Even if the ancillary input systems are entangled, they would correspond to a non-signaling process compatible with any causal configuration. Therefore, intuitively one would expect that adding such an input joint state to some parties embedded in a quantum causal process would result again to a causal process. We shown that this is not the case and refer to this effect as *activation* of non-causality.

We provide a particular example of a tripartite causal quantum process matrix, which we construct inspired by the non-causal bipartite process matrix presented in Ref. [2],

$$W^{A_I A_O B_I B_O} = \frac{1}{4}(\mathbb{1}^{A_I A_O B_I B_O} + \frac{1}{\sqrt{2}}\sigma_z^{A_I}\sigma_x^{B_I}\sigma_z^{B_O} + \frac{1}{\sqrt{2}}\sigma_z^{A_O}\sigma_z^{B_I}), \qquad (2.66)$$

which can violate a causal inequality. The input and output systems of all parties are two-level systems, except the input of Charlie which is trivial. The process we construct has the following form in the σ basis

$$W^{A_I A_O B_I B_O C_O} = \frac{1}{4}(\mathbb{1}^{A_I A_O B_I B_O C_O} + \frac{1}{\sqrt{2}}\sigma_z^{A_I}\sigma_z^{B_I}\sigma_z^{B_O}\sigma_x^{C_O} + \frac{1}{\sqrt{2}}\sigma_z^{A_O}\sigma_z^{B_I}\sigma_z^{C_O}).$$
$$(2.67)$$

This process matrix has the right normalization (Tr $W^{A_I A_O B_I B_O C_O} = d_{A_O}d_{B_O}d_{C_O}$), contains only the allowed type of terms (listed in Table 2.3) and is positive semidefinite. The latter can be seen through the fact that relative to the $\{|0\rangle, |1\rangle\}$ basis of the input system of B, B_I (which is the eigenbasis of σ_z with corresponding to eigenvalues $+1$ and -1 respectively, $\sigma_z = |0\rangle\langle 0| - |1\rangle\langle 1|$) we can write the process matrix as

$$W^{A_I A_O B_I B_O C_O} = |0\rangle\langle 0|^{B_I} \otimes \frac{1}{4}(\mathbb{1}^{A_I A_O B_O C_O} + \frac{1}{\sqrt{2}}\sigma_z^{A_I}\sigma_z^{B_O}\sigma_x^{C_O} + \frac{1}{\sqrt{2}}\sigma_z^{A_O}\sigma_z^{C_O}) +$$
$$(2.68)$$

$$|1\rangle\langle 1|^{B_I} \otimes \frac{1}{4}(\mathbb{1}^{A_I A_O B_O C_O} - \frac{1}{\sqrt{2}}\sigma_z^{A_I}\sigma_z^{B_O}\sigma_x^{C_O} - \frac{1}{\sqrt{2}}\sigma_z^{A_O}\sigma_z^{C_O}). \quad (2.69)$$

We notice that the operator $\frac{1}{4}(\mathbb{1}^{A_I A_O B_O C_O} + \frac{1}{\sqrt{2}}\sigma_z^{A_I}\sigma_z^{B_O}\sigma_x^{C_O} + \frac{1}{\sqrt{2}}\sigma_z^{A_O}\sigma_z^{C_O})$ is identical to that in Eq. (2.66) (except that in the place of B_I we have C_O) and that operator was shown to be positive semidefinite. The operator $\frac{1}{4}(\mathbb{1}^{A_I A_O B_O C_O} - \frac{1}{\sqrt{2}}\sigma_z^{A_I}\sigma_z^{B_O}\sigma_x^{C_O} - \frac{1}{\sqrt{2}}\sigma_z^{A_O}\sigma_z^{C_O})$ is the same as the one just discussed except that the nontrivial σ terms have a minus sign. This operator can be obtained from the first one by a unitary transformation (e.g., one that takes $\sigma_x^{C_O}$ to $-\sigma_x^{C_O}$ and $\sigma_z^{C_O}$ to $-\sigma_z^{C_O}$, such as $\sigma_y^{C_O}$).

This process matrix describes a causally separable process. To see this, notice that it permits no signaling from Alice and Bob to Charlie, and so it can be formally written as $\mathcal{W}^{A,B,C} = \mathcal{W}^{A,B|C} \circ \mathcal{W}^C$. Now see that conditionally on any operation of Charlie, which is most generally described by a CP map with a CJ operator $M^{C_O} \geq 0$ (remember that Charlie has a trivial input system which means that its dimension is one), Alice and Bob are left with a bipartite process with process matrix

$$W^{A_I A_O B_I B_O}_{M^{C_O}} = \mathrm{Tr}_{C_O}[(M^{C_O} \otimes \mathbb{1}^{A_I A_O B_I B_O}) W^{A_I A_O B_I B_O C_O}] / \mathrm{Tr}[M^{C_O}]. \qquad (2.70)$$

This process matrix is a linear combination of the identity and terms containing only σ_z operators on different subsystems, i.e., it is diagonal in a given local basis (the $\{|0\rangle, |1\rangle\}$ basis for each subsystem). It was shown in Ref. [2] that all such bipartite process matrices are causally separable (though we remark that the same was shown not to hold for multipartite processes [10]). Imagine now that we supply Bob and Charlie with the entangled input state $\frac{1}{2}|\Phi^+\rangle\langle\Phi^+|^{C_I'B_I'}$, which yields the new process

$$W^{A_I A_O B_I B_I' B_O C_I' C_O} = W^{A_I A_O B_I B_O C_O} \otimes \frac{|\Phi^+\rangle\langle\Phi^+|^{C_I'B_I'}}{2}. \tag{2.71}$$

If Charlie performs the identity unitary channel from C_I' to C_O in his laboratory (which means that he sends out his part of the entangled state), which is described by the CJ operator $M^{C_I'C_O} = |\Phi^+\rangle\langle\Phi^+|^{C_I'C_O}$, Alice and Bob are left with the bipartite process

$$W^{A_I A_O B_I B_I' B_O} = \frac{1}{4}(\mathbb{1}^{A_I A_O B_I B_I' B_O} + \frac{1}{\sqrt{2}}\sigma_z^{A_I}\sigma_z^{B_I}\sigma_x^{B_I'}\sigma_z^{B_O} + \frac{1}{\sqrt{2}}\sigma_z^{A_O}\sigma_z^{B_I}\sigma_z^{B_I'}). \tag{2.72}$$

To see this, notice that the fact that taking the partial trace of $W^{A_I A_O B_I B_I' B_O C_I' C_O}$ with the operator $|\Phi^+\rangle\langle\Phi^+|^{C_I'C_O}$ is formally identical (up to a normalization) to a local projection in a quantum-state teleportation protocol [40]. Remember that if we start with a prepared $|\Phi+\rangle$ state on a bipartite system, send one part of it to a party which performs a joint projective measurement on that part and another system which he possesses (all systems have same dimensions), then the state of the other part of the prepared entangled state is 'teleported' to the system the party has initially possessed. This means that the operation of Charlie amounts to 'teleporting' the part of the matrix on C_O, onto B_I'. Note that the notion of teleportation is defined for quantum states and not for process matrices, and that protocol requires a correcting operation on the receiver's side, as the above projection, which does not require a correction, cannot be accomplished deterministically [40]. Now see that the process matrix (2.72) is similar to (2.66), except that the local operators on B_I in the non-trivial sigma terms in Eq. (2.66) are now on B_I', and there is a σ_z operator on B_I in each such term.

This process matrix is non-causal, because it allows Alice and Bob to obtain any correlations that they could obtain using the non-causal process matrix (2.66). This can be done as follows. Alice always performs the same operations that she would perform with the process matrix (2.66). Bob performs a measurement on system B_I in the $\{|0\rangle, |1\rangle\}$ basis. If he obtains the outcome $|0\rangle$, then it is as if Alice and Bob share the process matrix (2.66) with B_I' in the place of B_I. He will then apply any operation from B_I' to B_O that he would apply from B_I to B_O with the process matrix (2.66), which yields the same joint probabilities for Alice and Bob as those with the process matrix (2.66). If Bob obtains the outcome $|1\rangle$ for his measurement on B_I, then it is as if Alice and Bob share the same process matrix as (2.66) with B_I' in the place of B_I but with a minus sign in front of each of the two nontrivial σ terms. This process matrix is equivalent to the previous one under a change of basis by the unitary $\sigma_y^{B_I'}$. Therefore, Bob can simply apply from B_I' to B_O the same operations he would apply from B_I to B_O with the process matrix (2.66) but transformed by the unitary transformation $\sigma_y^{B_1'}$. Again, this yields the same joint probabilities for Alice and Bob as with the process matrix (2.66). In particular, Alice and Bob can use this strategy to violate the causal inequality described in Ref. [2]. The process matrix (2.72) is thus non-causal, and so is the tripartite process matrix (2.71).

It is not known at present whether non-causal processes can be realized in agreement with the known laws of quantum mechanics without resorting to post-selection. We have seen in the previous subsection that we can realize causally non-separable processes, which are nevertheless causal. Here, we see that certain causal processes can become non-causal when supplied with shared entanglement. The ability to extend a process with shared entanglement seems natural to expect for any experimentally realizable process. From this perspective, this result suggests that either non-causal processes may be possible, or that there may exist causally separable processes, as defined above, that cannot be realized in practice.

2.4.5 Extensibly Causal and Extensibly Causally Separable Quantum Processes

Synopsis: Given the effect of activation of non-causality (the fact that some causal quantum processes can become non-causal by supplying the parties with entangled ancillas), we divide further the causal and causally separable quantum processes into those that can and cannot become non-causal, or causally non-separable respectively, by supplying the parties with entangled ancillas. The quantum processes that cannot be *activated* in this way are called extensibly causal and extensibly causally separable, respectively.

The activation of non-causality suggests that we distinguish those processes that can and cannot lose their property of causality by supplying the parties with entangled ancillas. Hence we propose the following definitions about the processes that retain their property of causality (by which we mean that they remain causal or remain causally separable) under any extension of their inputs.

Definition 2.4.2 (*Extensibly causal quantum process*) A quantum process that is causal and remains causal under extension with input systems in an arbitrary joint quantum state is called extensibly causal.

Definition 2.4.3 (*Extensibly causally separable (ECS) quantum process*) A quantum process that is causally separable and remains causally separable under extension with input systems in an arbitrary joint quantum state is called extensibly causally separable (ECS).

The process matrices of these types of processes will also be referred to as extensibly causal and ECS process matrices, respectively.

Note. These definitions can be formulated analogously for more general process theories that permit composite local systems.

The natural question to ask now is whether these classes of processes correspond to physical situations that we can describe, and whether these subsets constitute the whose set (for example are all causal processes extensibly causal?)

Observation 1: In the bipartite case, all causally separable processes are extensibly causally separable. To see this, think of adding an arbitrary joint input ancilla to a process matrix of the form (2.58); we obtain a process matrix of the same form. Therefore the notion of activation of causal non-separability can be thought to be a multipartite characteristic in the case of quantum processes.

Observation 2: Extensibly causal and extensibly causally separable processes are not equivalent in general. Indeed, the causally non-separable tripartite process (2.61) based on the quantum switch, which is causal, is also extensibly causal. This is because our proof that it is causal applies also if the parties share entangled input ancillas.

Comment: Recently, Feix, Araújo, and Brukner gave an example of a bipartite quantum process that is causal but not extensibly causal [41], proving that causality and extensible causality are different in the bipartite case too. While in the tripartite case we have seen that extensible causality is also different from causal separability, it is currently an open problem whether the same holds in the bipartite case.

In the next subsection, we derive a characterization of the tripartite ECS processes in terms of conditions on the form of the process matrix which generalize the conditions in the bipartite case (Eqs. (2.56), (2.58)).

2.4.6 Structure of Extensibly Causally Separable Process Matrices in the Tripartite Case

Synopsis: We derive necessary and sufficient conditions for a tripartite process to be extensibly causally separable. These are conditions on the form of the process matrix. They differ from the bipartite case as now the dynamical causal order plays a role. The result aimed to give way to a similar form for the general n-partite problem, although we found no simple extension. The obtained result for the tripartite case is also useful for the general characterization of this set of process matrices, for example when checking whether a given process matrix belongs to this set or when looking for so called *causal witnesses* (something that we discuss in the next Chapter). Similarly to finding an entanglement witness where it is necessary to define the space of separable states, when finding a causal witness we provide a definition for the space of extensibly causally separable process matrices.

For conveniency, we rewrite the definition of causally separable processes.

Definition 2.4.4 (*Causally separable quantum process*) A quantum process is called causally separable if and only if it can be decomposed in the canonical form given by Theorem 3 (which we re-write below for conveniency), with the additional condition that each process on the right-hand side of Eq. (2.73) is a quantum process. (Note that since the canonical form is defined iteratively, the latter is understood to hold for all conditional processes in this definition.)

Theorem (Canonical causal decomposition) *A causal process for n parties is one that can be written in the (generally non-unique) form*

$$\mathcal{W}_c^{1,\cdots,n} = \sum_{i=1}^{n} q_i \mathcal{W}^{(1,\cdots,i-1,i+1,\cdots,n)\not\preceq i}, \quad q_i \geq 0, \forall i, \ \sum_{i=1}^{n} q_i = 1, \tag{2.73}$$

with

$$\mathcal{W}^{(1,\cdots,i-1,i+1,\cdots,n)\not\preceq i} = \mathcal{W}_c^{1,\cdots,i-1,i+1,\cdots,n|i} \circ \mathcal{W}^i, \tag{2.74}$$

where the $(n-1)$*-partite conditional process* $\mathcal{W}_c^{1,\cdots,i-1,i+1,\cdots,n|i}$ *is either trivial (when* $n = 1$*) or has the form (2.73) for every value of the event in i.*

There is an immediate consequence of the definition of the causally separable process, for the structure of a causally separable process matrix. Given that the probabilities of a quantum process are linear in the process matrix, a causally separable process matrix can be written in the form

$$W_{cs}^{1_I 1_O \cdots n_I n_O} = \sum_{i=1}^{n} q_i W^{(1,\cdots,i-1,i+1,\cdots,n)\not\preceq i}, \quad 0 \leq q_i, \forall i, \ \sum_{i=1}^{n} q_i = 1, \tag{2.75}$$

where $W^{(1,\cdots,i-1,i+1,\cdots,n)\not\preceq i}$ is a process matrix which describes a process $\mathcal{W}^{(1,\cdots,i-1,i+1,\cdots,n)\not\preceq i}$ with the property

$$\mathcal{W}^{(1,\cdots,i-1,i+1,\cdots,n)\not\preceq i} = \mathcal{W}_{cs}^{1,\cdots,i-1,i+1,\cdots,n|i} \circ \mathcal{W}^i, \tag{2.76}$$

where for $n > 1$ the conditional process $\mathcal{W}_{cs}^{1,\cdots,i-1,i+1,\cdots,n|i}$ is a causally separable process for every value of the event in i, and for $n = 1$ it is the trivial process. Note that the requirement that $\mathcal{W}^{(1,\cdots,i-1,i+1,\cdots,n)\not\preceq i}$ is a quantum process that permits no signaling from the rest of the parties to i guarantees that both the reduced and the conditional process on the right-hand side of Eq. (2.76) are valid quantum processes (this can be seen from the no signaling condition in Proposition 2.4.2).

Aim: In the case of two parties (note that any monopartite process is trivially causally separable and ECS), we have seen that the process matrices $W^{A \not\preceq B}$, whose processes obey $\mathcal{W}^{A \not\preceq B} = \mathcal{W}_{cs}^{A|B} \circ \mathcal{W}^B$, are those that can be written in the form $W^{A \not\preceq B} = W^{B_I B_O A_I} \otimes \mathbb{1}^{A_O}$, and the general form of bipartite causally separable process matrices is (2.58). As noted already, this is also the general form of the bipartite ECS process matrices. Our goal is to obtain a similar condition for tripartite ECS processes.

Form of the $W^{(A,B)\not\preceq C}$: Let us consider a process of the form $\mathcal{W}^{(A,B)\not\preceq C} = \mathcal{W}_{cs}^{A,B|C} \circ \mathcal{W}^C$, where $\mathcal{W}^C$ is a monopartite quantum process and $\mathcal{W}_{cs}^{A,B|C}$ is a bipartite conditional process which is causally separable for each possible event in C. Notice that since there should be no signaling from Alice and Bob to Charlie in such a process, its process matrix, which we will denote $W_{(A,B)\not\preceq C}^{A_I A_O B_I B_O C_I C_O}$, can at most contain the types of terms listed in Table 2.4. These are the terms that do not permit signaling from Alice and Bob to Charlie according to Proposition 2.4.2.

We will first obtain necessary and sufficient conditions for such a process $(\mathcal{W}^{(A,B)\not\preceq C})$ to be ECS. Note that we have not proven yet that a general ECS process matrix should have the form (2.75) where each of the terms $W^{(1,\cdots,i-1,i+1,\cdots,n)\not\preceq i}$ is itself ECS. This will be shown later.

Every event in Charlie's laboratory is described by some CP map with CJ operator $M^{C_I C_O} \geq 0$, $\mathrm{Tr}\, M^{C_I C_O} \leq d_{C_I}$. Conditionally on such an event, Alice and Bob are left with the process matrix

$$W_{M^{C_I C_O}}^{A_I A_O B_I B_O} = \mathrm{Tr}_{C_I C_O}[W_{(A,B)\not\preceq C}^{A_I A_O B_I B_O C_I C_O}(\mathbb{1}^{A_I A_O B_I B_O} \otimes M^{C_I C_O})]/p(M^{C_I C_O}), \quad (2.77)$$

where $p(M^{C_I C_O})$ is the probability for the event $M^{C_I C_O}$ to occur in Carlie's laboratory (given the appropriate setting), which is independent of the operations performed by Alice and Bob since the process involves no signaling from Alice and Bob to Charlie. More specifically,

$$p(M^{C_I C_O}) = \mathrm{Tr}[W^{C_I C_O} M^{C_I C_O}], \quad (2.78)$$

where

$$W^{C_I C_O} = \mathrm{Tr}_{A_I A_O B_I B_O}[W_{(A,B)\not\preceq C}^{A_I A_O B_I B_O C_I C_O}(\frac{\mathbb{1}^{A_I A_O B_I B_O}}{d_{A_O} d_{B_O}} \otimes \mathbb{1}^{C_I C_O})] \quad (2.79)$$

is the reduced process of Charlie. The requirement that the conditional process for Alice and Bob is causally separable means that for all $M^{C_I C_O}$,

$$W_{M^{C_I C_O}}^{A_I A_O B_I B_O} = q_{M^{C_I C_O}} W_{M^{C_I C_O}}^{A\not\preceq B} + (1 - q_{M^{C_I C_O}})W_{M^{C_I C_O}}^{B\not\preceq A}, \quad (2.80)$$

where $W_{M^{C_I C_O}}^{A\not\preceq B}$ and $W_{M^{C_I C_O}}^{B\not\preceq A}$ are valid process matrices compatible with $A \not\preceq B$ and $B \not\preceq A$, respectively, and $q_{M^{C_I C_O}} \in [0, 1]$ (all objects generally depend on $M^{C_I C_O}$). For convenience, we will write this simply in the form

$$W_{M^{C_I C_O}}^{A_I A_O B_I B_O} = \mathbb{1}^{A_O} \otimes \tilde{W}_{M^{C_I C_O}}^{A_I B_I B_O} + \mathbb{1}^{B_O} \otimes \tilde{W}_{M^{C_I C_O}}^{A_I A_O B_I}, \quad (2.81)$$

where $\tilde{W}_{M^{C_I C_O}}^{A_I B_I B_O} \geq 0$ and $\tilde{W}_{M^{C_I C_O}}^{A_I A_O B_I} \geq 0$, and the whole operator is a valid process matrix, i.e., it contains only allowed terms and is properly normalized.

Initial result (sufficiency): A sufficient condition for this to hold is that

$$W^{A_I A_O B_I B_O C_I C_O}_{(A,B) \not\preceq C} = \mathbb{1}^{A_O} \otimes \tilde{W}^{A_I B_I B_O C_I C_O} + \mathbb{1}^{B_O} \otimes \tilde{W}^{A_I A_O B_I C_I C_O}, \qquad (2.82)$$

where $\tilde{W}^{A_I B_I B_O C_I C_O} \geq 0$ and $\tilde{W}^{A_I A_O B_I C_I C_O} \geq 0$ are some positive semidefinite operators, whose sum gives a properly normalized quantum process matrix containing only the types of terms listed in Table 2.4. An important remark is that each of $\tilde{W}^{A_I B_I B_O C_I C_O} \geq 0$ and $\tilde{W}^{A_I A_O B_I C_I C_O} \geq 0$ may contain terms that are forbidden in a process matrix, such as terms of type C_O, but these terms have to cancel in the sum. Indeed, we have

$$\mathrm{Tr}_{C_I C_O}[W^{A_I A_O B_I B_O C_I C_O}_{(A,B) \not\preceq C}(\mathbb{1}^{A_I A_O B_I B_O} \otimes M^{C_I C_O})]/p(M^{C_I C_O}) =$$
$$W^{A_I A_O B_I B_O}_{M^{C_I C_O}} = \mathbb{1}^{A_O} \otimes \tilde{W}^{A_I B_I B_O}_{M^{C_I C_O}} + \mathbb{1}^{B_O} \otimes \tilde{W}^{A_I A_O B_I}_{M^{C_I C_O}}, \qquad (2.83)$$
$$\forall M^{C_I C_O} \geq 0,$$

where

$$\tilde{W}^{A_I B_I B_O}_{M^{C_I C_O}} = \mathrm{Tr}_{C_I C_O}[\tilde{W}^{A_I B_I B_O C_I C_O}(\mathbb{1}^{A_I B_I B_O} \otimes M^{C_I C_O})]/p(M^{C_I C_O}) \geq 0, \quad (2.84)$$

$$\tilde{W}^{A_I A_O B_I}_{M^{C_I C_O}} = \mathrm{Tr}_{C_I C_O}[\tilde{W}^{A_I A_O B_I C_I C_O}(\mathbb{1}^{A_I A_O B_I} \otimes M^{C_I C_O})]/p(M^{C_I C_O}) \geq 0, \quad (2.85)$$

and it is easy to see that since $W^{A_I A_O B_I B_O C_I C_O}_{(A,B) \not\preceq C}$ contains only the types of terms listed in Table 2.4, $W^{A_I A_O B_I B_O}_{M^{C_I C_O}}$ can only contain allowed terms. To verify this, notice that if a process matrix has the types of terms listed in Table 2.4, then for every CP map of Charlie (that has at most terms of the type $C_I, C_O, C_I C_O$) the remaining process will have the terms that are of the same type as in that Table, but with $C_I, C_O, C_I C_O$ removed. The remaining terms are those allowed in a valid bipartite process matrix.

The above analysis is for the process matrix of a process of the form $\mathcal{W}^{(A,B) \not\preceq C} = \mathcal{W}^{A,B|C}_{cs} \circ \mathcal{W}^{C}$, which is by definition a causally separable process, and hence the process matrix is causally separable. It is immediate to see that this condition is sufficient also for the process matrix $W^{A_I A_O B_I B_O C_I C_O}_{(A,B) \not\preceq C}$ to be ECS. This is because if $W^{A_I A_O B_I B_O C_I C_O}$ has the above properties, any extension $W^{A_I A_O B_I B_O C_I C_O} \otimes \rho^{A'_I B'_I C'_I}$, where $\rho^{A'_I B'_I C'_I}$ is a density matrix, also has these properties.

Initial result (necessity): We now show that the form (2.82) is also a necessary condition for an ECS process matrix compatible with $(A, B) \not\preceq C$, which we will denote by $W^{A_I A_O B_I B_O C_I C_O}_{ecs;(A,B) \not\preceq C}$. To see this, imagine that we supply Alice and Charlie respectively with ancillary systems A'_I and C'_I of dimension $d_{C_1} d_{C_2}$ each, which are prepared in the maximally entangled state $|\phi^+\rangle\langle\phi^+|^{A'_I C'_I}/(d_{C_1} d_{C_2})$, where $|\phi^+\rangle = \sum_{i=1}^{d_{C_1} d_{C_2}} |i\rangle^{A'_I}|i\rangle^{C'_I}$. Conditionally on Charlie performing a suitable operation and obtaining an outcome with CP map $M^{C_I C_O C'_I} \propto |\phi^+\rangle\langle\phi^+|^{(C_I C_O)C'_I}$, Alice and Bob will be left sharing a process matrix which, up to a normalization factor, has an identical form to that of $W^{A_IA_OB_IB_OC_IC_O}_{ecs;(A,B)\not\preceq C}$ but with A'_I in the place of C_IC_O. The requirement that this is a causally separable bipartite process matrix means that $W^{A_IA_OB_IB_OC_IC_O}_{ecs;(A,B)\not\preceq C}$ must be of the form (2.82).

Final result: So far, we have only obtained necessary and sufficient conditions for an ECS process matrix $W^{A_IA_OB_IB_OC_IC_O}_{ecs;(A,B)\not\preceq C}$ compatible with $(A,B)\not\preceq C$ (and similarly for permutations of A, B, C). We next prove the general case.

Proposition 2.4.3 *Every tripartite ECS process matrix can be written in the form*

$$W^{A_IA_OB_IB_OC_IC_O}_{ecs} = q_1 W^{A_IA_OB_IB_OC_IC_O}_{ecs;(A,B)\not\preceq C} + q_2 W^{A_IA_OB_IB_OC_IC_O}_{ecs;(A,C)\not\preceq B} + q_3 W^{A_IA_OB_IB_OC_IC_O}_{ecs;(B,C)\not\preceq A},$$

$$q_i \geq 0, \ \forall i = 1, 2, 3, \ \sum_{i=1}^{3} q_i = 1,$$

(2.86)

where $W^{A_IA_OB_IB_OC_IC_O}_{ecs;(A,B)\not\preceq C}$ contains only terms from Table 2.4 and has the form (2.82), and analogously for $W^{A_IA_OB_IB_OC_IC_O}_{ecs;(A,C)\not\preceq B}$ and $W^{A_IA_OB_IB_OC_IC_O}_{ecs;(B,C)\not\preceq A}$ by permutation.

The fact that this form is sufficient for the process matrix to be ECS is obvious because if this is true for each of the individual terms, any extension $W^{A_IA_OB_IB_OC_IC_O}_{ecs}$ $\otimes \rho^{A'_IB'_IC'_I} = q_1 W^{A_IA_OB_IB_OC_IC_O}_{ecs;(A,B)\not\preceq C} \otimes \rho^{A'_IB'_IC'_I} + q_2 W^{A_IA_OB_IB_OC_IC_O}_{ecs;(A,C)\not\preceq B} \otimes \rho^{A'_IB'_IC'_I} +$ $q_3 W^{A_IA_OB_IB_OC_IC_O}_{ecs;(B,C)\not\preceq A} \otimes \rho^{A'_IB'_IC'_I}$ is also causally separable. The fact that it is necessary can be seen as follows. Let us choose $\rho^{A'_IB'_IC'_I}$ which is a tensor product of three bipartite maximally entangled states of the type used in the 'teleportation' argument, one shared between Alice and Bob, the other one between Alice and Charlie, and the third one between Bob and Charlie. For this particular ancilla, it must be possible to write the extended process in the form

$$W^{A_IA_OB_IB_OC_IC_O}_{ecs} \otimes \rho^{A'_IB'_IC'_I} = q_1 W^{A_IA_OB_IB_OC_IC_O}_1 \otimes \rho^{A'_IB'_IC'_I}$$
$$+ q_2 W^{A_IA_OB_IB_OC_IC_O}_2 \otimes \rho^{A'_IB'_IC'_I} + q_3 W^{A_IA_OB_IB_OC_IC_O}_3 \otimes \rho^{A'_IB'_IC'_I},$$

(2.87)

where $W^{A_IA_OB_IB_OC_IC_O}_1 \otimes \rho^{A'_IB'_IC'_I}$ is causally separable and compatible with $(A,B)\not\preceq C$, $W^{A_I\cdots C_O}_2 \otimes \rho^{A'_IB'_IC'_I}$ is causally separable and compatible with $(A,C)\not\preceq B$, and $W^{A_I\cdots C_O}_3 \otimes \rho^{A'_IB'_IC'_I}$ is causally separable and compatible with $(B,C)\not\preceq A$. But for each of these terms, we can perform the 'teleportation' argument exploiting the respective maximally entangled bipartite state contained in $\rho^{A'_IB'_IC'_I}$, proving that $W^{A_I\cdots C_O}_1$ has the form we obtained for $W^{A_I\cdots C_O}_{ecs;(A,B)\not\preceq C}$, $W^{A_I\cdots C_O}_2$ has the form we obtained for $W^{A_I\cdots C_O}_{ecs;(A,C)\not\preceq B}$, and $W^{A_I\cdots C_O}_3$ has the form we obtained for $W^{A_I\cdots C_O}_{ecs;(B,C)\not\preceq A}$. This completes the proof.

2.4.7 Processes Realizable by Classically Controlled Quantum Circuits

Synopsis: In this section we find a class of extensibly causally separable (ECS) processes that are physically realizable. We define a general model in which parties perform black-box operations. A prior measurement determines which party is first. After the operation of the first party, their outcome and the outcome of a subsequent measurement by the circuit determines which party is second, and so on. We make two claims: any ECS process can be implemented by this model and that any process obtained by this model is ECS. We show that both claims hold in the bipartite case. In the multipartite case, we show that the second claim is true and provide good reasons to conjecture that the first claim is also true.

In this section we are interested in a class of processes that have a physical realization and show its relation with the class of extensibly causally separable processes (ECS). In the bipartite case the experimental realization of an ECS process is trivial: most generally it will be a probabilistic mixture of causally-ordered quantum circuits for two parties. For the multipartite case, the answer is not simple. Here we show that a particular class of processes which can be realized in practice, referred to as classically controlled quantum circuits, belong to the class of ECS processes and we conjecture that it in fact gives rise to all ECS processes (this is true in the bipartite case).

The general protocol: The main idea our considered process should capture is the fact that the outcome of some operations can affect the order in which future operations occur. We define such a process to have the following general realization. We consider a number of parties N, whose operations can be thought of, most generally, as black-box operations. In that case, although we may not have access to measurement outcomes of those black-box operations, it is nonetheless possible that the order of subsequent operations in the circuit may depend indirectly on the event inside the black box. This is because the order can be decided based on a measurement on the output system.

We begin with some sufficiently large quantum system, or register, in a given quantum state. We perform a quantum operation on it and conditionally on the outcome we determine which party will be first, which subsystem of the register will be the input of that party, and what operation will be applied after that party. These decisions will be according to some rule that has been specified before the experiment. We then apply the black-box operation of the first party on the decided subsystem, perform the decided operation after it, and conditionally on that outcome and the outcome of the party it is decided which party will be second, and so on. This continues until all N parties have been called (only once). This model gives rise to valid quantum processes, as we can formally write the operation of each box and calculate the joint probabilities for the outcomes of all boxes using standard quantum mechanics and see that they are linear and non-contextual functions of the respective CP maps of the parties. The same holds if we introduce ancillary systems prepared in an arbitrary quantum state and allow the parties to perform their operations in the extended system of their inputs and parts of them.

Aim: We are interested in the following two claims, which we investigate in the bipartite and multipartite case: any ECS process can be implemented by this model, and any process obtained by this model is ECS.

Bipartite case (both claims): In the case of two parties, we know that any (extensibly) causally separable process can be implemented in this way: it most generally corresponds to embedding at random the local experiments of the parties, A and B, into one of two possible fixed circuits (with alternate order of the parties), which can be chosen conditionally on the outcome of a measurement on some state at the very beginning of the experiment. That measurement alone is enough to determine the total order of the parties, as once the first party is determined the second will just have to be the other party. Therefore the first claim holds. To see that the second claim also holds, notice first that the process is independent of the operation applied after the last party (there are no more parties whose order might be conditioned on that operation or any other succeeding operation). Also, the outcome of any operation after the first party can be ignored since there is only one choice of the last party, i.e. that operation can be assumed to be deterministic. Finally, the outcomes of the operation before the first party (the ones that determine who is first and therefore which fixed circuit will be implemented) can be grouped into two coarse-grained outcomes (one for each fixed circuit). Therefore, the process realized by such a procedure is a probabilistic mixture of the processes of two fixed-order circuits, which is the claimed form.

Multipartite case (claim 2): In the case of more than two parties, we can easily show that the second claim holds. First notice that depending on the outcome of the first measurement (whose probability is independent of any future operations and therefore the settings of the parties), there will be one party that is first and hence the subsequent process that results from the process that results from the protocol can involve no-signaling from the rest of the parties to that first party. Therefore, there is a well-defined reduced process for the first party. Taking into account all possible outcomes of the first measurement, the whole process will be a probabilistic mixture of processes where one party is first and there is a well-defined reduced process for that party. This is what Eq. (2.75) describes. Now conditionally on the outcome of the first party, the procedure for the rest of the parties looks analogous, so Eq. (2.76) holds too, i.e. the process is causally separable. The argument remains even if the parties are supplied with ancillas in any quantum state, and therefore, any process realized by classically controlled quantum circuits is extensibly causally separable.

Multipartite case (claim 1): We conjecture that the first claim also holds. We provide some partial considerations that support this conjecture, based on the restrictions on the allowed terms in the process matrix (that gives rise to processes) realized by classically controlled quantum circuits in the tripartite case. We will focus on the question of implementing an ECS process matrix of the type $W_{ecs}^{(A,B)\npreceq C}$, which has the form (2.82) (compatible with $(A, B) \npreceq C$), by a classically controlled quantum

Fig. 2.9 Realization of an
ECS process compatible with
$(A, B) \not\preceq C$ by a classically
controlled quantum circuit

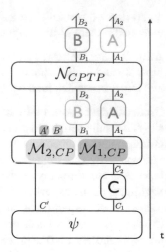

circuit. Implementability of a process matrix of this kind is both necessary and sufficient for the implementability of a general tripartite ECS process matrix (mixtures of different $(X, Y) \not\preceq Z$ as described in Proposition 2.4.3, since by using a suitable measurement at the beginning we can select with the right probability which of the three process matrices in the mixture on the right-hand side of Eq. (2.86) to realize subsequently.

The protocol: The protocol, depicted in Fig. 2.9, begins with some quantum system prepared in a state ρ. After Charlie operates on some subsystem C_1, we apply some operation, say $\mathcal{M}$ and conditionally on the outcome we determine who is second, on what subsystem they act, and what operation will be applied after that, say $\mathcal{N}$. Note that without loss of generality we can assume that the subsystem on which the second party acts is specified from the beginning. This is because any subsystem of the same dimension can be mapped onto the designated subsystem by a unitary transformation that can be part of the definition of the present operation just before that second party. Also, without loss of generality we can assume that this operation ($\mathcal{M}$) has only two outcomes, since we can group the outcomes into those for which Alice would be next, and those for which Bob will be next. Also, we can assume that the operation that takes place after the next party ($\mathcal{N}$), conditionally on the fine-grained outcome of the operation before that party ($\mathcal{M}$) within the same group of outcomes, will always be the same one for the one group of outcomes, and another one for the other group of outcomes. That is, conditionally on the outcome of the operation that takes place after Charlie ($\mathcal{M}$), the next party is determined, and the operation after that party is also determined ($\mathcal{N}$), that is performed on the combination of the output system of that party and some subsystem on which the classical information about the outcome of the operation (before that party) is copied. This is something that we can include as part of the definition of the operation after Charlie ($\mathcal{M}$). After occurrence of the operation after the second party ($\mathcal{N}$), there is only one possibility for the last party, and so the operation after the second party can be regarded as deterministic (as a

CPTP map) from all systems to the input of the last party. However, we do leave the possibility that this last operation ($\mathcal{N}$) may be defined conditionally on the first outcome (of $\mathcal{M}$), rather than absorb the conditioning on that outcome into a larger operation. This is to avoid complications arising from the fact that the different parties may have input and output systems of different dimensions. The outlined procedure is sketched in Fig. 2.9, where the two possible sequences of transformations arising from the two possible outcomes of the first operation ($\mathcal{M}$) are depicted in blue and green respectively. The two CP maps corresponding to the outcomes of the operation $\mathcal{M}$ must sum up to a CPTP map, since they correspond to the two possible outcomes of a standard quantum operation.

The protocol as process matrices: We remind that any process realized by this protocol is extensibly causally separable and we conjecture that any extensibly causally separable process can be realized by this protocol. Each of the two possible developments, depicted as blue and green maps and parties, is a non-deterministic linear supermap [46], that maps the local CP maps of the parties into the real numbers, which is the probability for the particular sequence of events. This can be written in a similar form as the formula for the probabilities of the outcomes of the parties in a valid process, except that in the place of the process matrix we would have an operator $\tilde{W}_1^{A_I A_O B_I B_O C_I C_O} \geq 0$ and $\tilde{W}_2^{A_I A_O B_I B_O C_I C_O} \geq 0$ for each possible development of the protocol, which generally would not be a valid process matrix. However, their sum, $\tilde{W}_1^{A_I A_O B_I B_O C_I C_O} + \tilde{W}_2^{A_I A_O B_I B_O C_I C_O} = W_{cs;(A,B)\not\leq C}^{A_I A_O B_I B_O C_I C_O}$, would be a valid process matrix realized through this classically controlled quantum circuit.

Core of the conjecture: For a causally separable process matrix, $W_{cs;(A,B)\not\leq C}^{A_I A_O B_I B_O C_I C_O}$, the operators in the sum $\tilde{W}_1^{A_I A_O B_I B_O C_I C_O} + \tilde{W}_2^{A_I A_O B_I B_O C_I C_O} = W_{cs;(A,B)\not\leq C}^{A_I A_O B_I B_O C_I C_O}$, obey certain restrictions in the type of non-trivial terms they can contain, that arise from the restriction that $W_{cs;(A,B)\not\leq C}^{A_I A_O B_I B_O C_I C_O}$ is a valid process matrix. We will show that the operator arising from each of the two possible developments of the protocol obeys the same restrictions on the types of allowed non-trivial terms. This suggests that each possible development *might* be described precisely by each of those terms, which suggests that the conjecture that *any* ECS process matrix can be realized by a classically controlled quantum circuit *could be* true.

Consider now just one of the two possible developments, say, the blue one, in which Alice is second and Bob is last (labeled by 1). Since Bob is last and his output system is discarded, we have $\tilde{W}_1^{A_I A_O B_I B_O C_I C_O} = \mathbb{1}^{B_O} \otimes \tilde{W}_1^{A_I A_O B_I C_I C_O}$ (similarly, in the other case we have $\tilde{W}_2^{A_I A_O B_I B_O C_I C_O} = \mathbb{1}^{A_O} \otimes \tilde{W}_2^{A_I B_I B_O C_I C_O}$). Notice that if the transformation $\mathcal{N}_{1,CPTP}$ after Alice was not required to be CPTP but could be any CP map for a suitable choice of the initial state ρ and of the CP maps $\mathcal{M}_{1,CP}$ and $\mathcal{N}_{1,CP}$ we could realize any $\tilde{W}_1^{A_I A_O B_I C_I C_O} \geq 0$. This means that there would be no restrictions on the types of non-trivial terms it contains. To see this, notice that we can choose the density operator $\rho^{C_I C'}$ proportional to $\tilde{W}_1^{A_I A_O B_I C_I C_O}$, where the part of $\tilde{W}_1^{A_I A_O B_I C_I C_O}$ on $A_I A_O B_I C_O$ is stored on C'. Then we can 'teleport' this part of the operator onto its desired subsystem by using CP maps $\mathcal{M}_{1,CP}$ and $\mathcal{N}_{1,CP}$ that have CJ operators proportional to projectors on maximally entangled states as needed to realize the 'teleportation' (the traces of these CP maps can be chosen to ensure the overall trace of the resultant operator $\tilde{W}_1^{A_IA_OB_IC_ICO}$). However, we cannot realize this scheme if the transformation after Alice is trace-preserving, $\mathcal{N}_{1,CPTP}$.

The particular constraints on what kind of $\tilde{W}_1^{A_IA_OB_IC_IC_O}$ can be obtained, given that the transformation after Alice is trace-preserving, lies on the fact that the CJ operator of $\mathcal{N}_{1,CPTP}$ cannot contain terms of type A_O, $A'A_O$, and A' (non-trivial terms on the input of the map). Considering the calculation of $\tilde{W}_1^{A_IA_OB_IC_IC_O}$ based on the CJ operators of ρ, $\mathcal{M}_{1,CP}$ and $\mathcal{N}_{1,CPTP}$, we see that the lack of these types of terms in $\mathcal{N}_{1,CPTP}$ implies the lack of any term with a nontrivial σ on A_O in $\tilde{W}_1^{A_IA_OB_IC_IC_O}$. This is the only constraint on the possible types of terms in $\tilde{W}_1^{A_IA_OB_IC_IC_O}$. These allowed types of terms are exactly those allowed in the operator $\tilde{W}^{A_IA_OB_IC_IC_O}$ in Eq. (2.82). Similarly, we see that the allowed terms in $\tilde{W}_2^{A_IB_IB_OC_IC_O}$ (Bob second, Alice last) are the same as those in $\tilde{W}^{A_IB_IB_OC_IC_O}$ in Eq. (2.82). These are the terms allowed in a process matrix compatible with Charlie being first, except that both $\tilde{W}_1^{A_IA_OB_IC_IC_O}$ and $\tilde{W}_2^{A_IB_IB_OC_IC_O}$ may contain terms of type C_O. The fact that these terms should cancel in the sum $\mathbb{1}^{B_O} \otimes \tilde{W}_1^{A_IA_OB_IC_IC_O} + \mathbb{1}^{A_O} \otimes \tilde{W}_2^{A_IB_IB_OC_IC_O} = W_{cs;(A,B)\not\prec C}^{A_IA_OB_IB_OC_IC_O}$ follows from the fact that this is a valid ECS process, and can be seen to be ensured by the requirement that $\mathcal{M}_{1,CP} + \mathcal{M}_{2,CP}$ is CPTP.

The only restriction on the operators $\mathbb{1}^{B_O} \otimes \tilde{W}_1^{A_IA_OB_IC_IC_O}$ and $\mathbb{1}^{A_O} \otimes \tilde{W}_2^{A_IB_IB_OC_IC_O}$ imposed by this model, apart from their positive-semidefiniteness and the normalization of their sum, seems to be the absence of the forbidden terms in each of them, as well as of the forbidden terms in their sum. If this is indeed the case, then any ECS process could be realized by a suitable classically controlled quantum circuit. A strictly rigorous proof requires showing that apart from the lack of these forbidden terms, there can be no other hidden constraints on the pair of operators $\mathbb{1}^{B_O} \otimes \tilde{W}_1^{A_IA_OB_IC_IC_O}$ and $\mathbb{1}^{A_O} \otimes \tilde{W}_2^{A_IB_IB_OC_IC_O}$ (which, of course, are guaranteed to be properly normalized). One way of doing it could be by exhibiting an explicit constructive procedure for implementing any given ECS process, which would be of additional interest on its own right.

2.5 Conclusion

Main result: In this chapter, we proposed a rigorous definition of causality in the process framework [2]. This definition is the first one to take into account the fact that the causal order between a set of local experiments may be random and correlated with the settings of some of them.

Multipartite causal processes: Multipartite processes that obey our definition of causality, referred to as causal, and thus permitting such 'dynamical' causal order, have a particular structure which we derived. It is an iteratively formulated canonical form expressed in terms of reduced and conditional processes (well-defined processes conditioned on future and past events respectively). Specifically, a causal process is one that has a causal decomposition in which each term is a process compatible with one party being first; the latter can be written as a multiplication of a monopartite process of the first party, and a conditional process of the rest of the parties, given the first. This conditional process has to be also causal. This form can be interpreted as an unraveling of the process into a sequence of local experiments, which agrees with the condition that the order and outcomes of the experiments prior to a given step is independent of the settings of future experiments. Defining causal processes serves as a tool to investigate non-causal processes in a theory independent scenario, as the probabilities of a causal process form a polytope whose facets define causal inequalities. Their violation by a given process can be interpreted as demonstrating the non-existence of causal order between the local experiments.

Quantum case: multipartite causal and causally separable processes: In the quantum process framework, where the local experiments are described by standard quantum mechanics, we investigated the relationship between two concepts: causality and causal separability. The latter concept was introduced in Ref. [2] for the bipartite case. We proposed a definition of causal separability for the multipartite case, which reduces to the one for the case of two parties, based on the canonical form of causal processes. Specifically, we defined a causally separable quantum process as a causal quantum process that has a causal decomposition such that the different processes in each term of the decomposition are themselves valid quantum processes. We showed the set of causally separable quantum processes is strictly within the set of causal quantum processes, by exhibiting an example of a tripartite process that is causal but not causally separable. Recently, the same was shown to hold also in the bipartite case [41]. We also gave an example of a causally separable (and hence also causal) process that becomes non-causal when extended by supplying the parties with an entangled ancillary state. Based on this observation, we proposed two extended notions of causality and causal separability called extensible causality and extensible causal separability, which require preservation of the respective property under extending the process with entangled input ancillas. Although they are different in the general case, the sets of causally separable and ECS processes are equivalent in the bipartite case. We showed that the sets of extensibly causal and causally separable processes are different in general via the same tripartite example that we used to show that causal and causally separable processes are different. We did not investigate whether the same separation holds in the bipartite case. However, it was recently shown that causal and extensibly causal processes are different in the bipartite case too [41].

Results on extensibly causally separable processes: Finally, we derived a simple characterization of the ECS quantum processes in the tripartite case in terms of conditions on the form of their process matrices, which extends the conditions for (extensibly) causally separable process matrices in the bipartite case. We conjectured that the set of ECS processes is equivalent to the processes that can be obtained within the paradigm of classically controlled quantum circuits and provided evidence for this based on analysis of the restrictions that this paradigm imposes on the tripartite

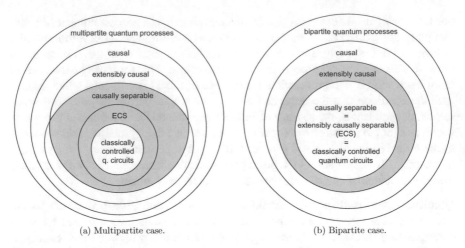

(a) Multipartite case.   (b) Bipartite case.

Fig. 2.10 A Venn diagrammatic sketch of our present knowledge of the different sets of quantum processes that we have introduced, in the general multipartite case and in the bipartite case. The white segments are non-empty. The gray segments are sets for which at present we do not know if they are empty or not [1]

process matrices it can create. The ECS processes and the processes obtainable by classically controlled quantum circuits are equivalent in the bipartite case.

Summary on the classification of processes: Our present understanding of the relation between all these different classes of quantum processes is illustrated for the general multipartite case and for the bipartite case in Fig. 2.10a and b, respectively. An obvious open problem is whether the gray segments in these figures are empty or not.

Open questions: Another problem of fundamental importance is to understand the class of quantum processes that are physically admissible in agreement with the known laws of quantum mechanics (see [48] for a relevant claim), and where this class stands with respect to all of the above classes. Are the processes that can be realized by classically controlled quantum circuits all the physically admissible causally separable processes? Where does the class of quantum-controlled quantum circuits stand? At present, this is the most general operationally feasible paradigm that we are aware of and all known processes realizable through it seem to be extensibly causal. Could the class of extensibly causal processes be equivalent to quantum-controlled quantum circuits? Did the reader really read the whole chapter? And most intriguingly, are there physically admissible non-causal processes?

The implications of our results are not limited to the subject of indefinite causal order in quantum mechanics. They can be useful also for the problem of inferring causal structure [28], both in classical and quantum theory [47]. The subject of causal inference concerns many disciplines, from philosophy and machine learning to sociology and medicine. Our formulation of a background-independent operational notion of causality that admits dynamical causal relations opens the road to a more general paradigm for causal inference than the one assuming deterministic underlying variables and static causal relations [28]. The decomposition of causal processes derived here implies constraints on the possible causal orders compatible with given setting-outcome correlations. This can serve as a basis for developing more sophisticated tools for causal inference for the case of dynamical causal order for an arbitrary number of parties.

References

1. Oreshkov O, Giarmatzi C (2016) New J Phys 18:093020
2. Oreshkov O, Costa F, Brukner Č (2012) Quantum correlations with no causal order. Nat Commun 3:1092
3. Hardy L (2005) Probability theories with dynamic causal structure: a new framework for quantum gravity. arXiv:0509120
4. Hardy L (2007) Quantum gravity computers: on the theory of computation with indefinite causal structure. arXiv:quant-ph/0701019
5. Chiribella G, D'Ariano GM, Perinotti P, Valiron, B (2013) Beyond causally ordered quantum computers. Phys Rev A 88, 022318. arXiv:0912.0195 (2009)
6. Colnaghi T, D'Ariano GM, Perinotti P, Facchini S (2012) Quantum computation with programmable connections between gates. Phys Lett A 376:2940–2943
7. Chiribella G (2012) Perfect discrimination of no-signalling channels via quantum superposition of causal structures. Phys Rev A 86:040301
8. Baumeler Ä, Wolf S (2014) Perfect signaling among three parties violating predefined causal order. Proc Int Symp Inf Theory (ISIT) 2014:526–530
9. Nakago K, Hajdušek M, Nakayama S, Murao M (2015) Parallelized adiabatic gate teleportation. Phys Rev A 92:062316
10. Baumeler Ä, Feix A, Wolf S (2014) Maximal incompatibility of locally classical behavior and global causal order in multi-party scenarios. Phys Rev A 90:042106
11. Araújo M, Costa F, Brukner Č (2014) Computational advantage of quantum-controlled ordering of gates. Phys Rev Lett 113:250402
12. Brukner Č (2014) Quantum causality. Nat Phys 10(4):259–263
13. Morimae T (2014) Acausal measurement-based quantum computing. Phys Rev A 90:010101(R)
14. Ibnouhsein I, Grinbaum A (2015) Information-theoretic constraints on correlations with indefinite causal order. Phys Rev A 92:042124
15. Brukner Č (2015) Bounding quantum correlations with indefinite causal order. New J Phys 17:073020
16. Oreshkov O, Cerf NJ (2016) Operational quantum theory without predefined time. New J Phys 18:073037. http://stacks.iop.org/1367-2630/18/i=7/a=073037
17. Oreshkov O, Cerf NJ (2015) Operational formulation of time reversal in quantum theory. Nat Phys 11:853–858
18. Procopio LM et al (2015) Experimental superposition of orders of quantum gates. Nat Commun 6:7913. https://doi.org/10.1038/ncomms8913
19. Rubino G et al (2016) Experimental verification of an indefinite causal order. Sci Adv 2. http://advances.sciencemag.org/content/3/3/e1602589.abstract
20. Lee CM, Barrett J (2015) Computation in generalised probabilistic theories. New J Phys 17:083001. http://stacks.iop.org/1367-2630/17/i=8/a=083001
21. Araújo M, Branciard C, Costa F, Feix A, Giarmatzi C, Brukner Č (2015) Witnessing causal nonseparability. New J Phys 17:102001

22. Baumeler Ä, Wolf S (2016) The space of logically consistent classical processes without causal order. New J Phys 18:013036. http://stacks.iop.org/1367-2630/18/i=1/a=013036
23. Branciard C, Araújo M, Feix A, Costa F, Brukner Č (2016) The simplest causal inequalities and their violation. New J Phys 18:013008
24. Abbott AA, Giarmatzi C, Costa F, Branciard C (2016) Multipartite causal correlations: polytopes and inequalities. Phys Rev A 94:032131
25. Giarmatzi C, Costa F (2018) A quantum causal discovery algorithm. npj Quantum Inf 4:17
26. Portmann C, Matt C, Maurer U, Renner R, Tackmann B (2017) Causal boxes: quantum information-processing systems closed under composition. IEEE Trans Inf Theory 63(5):3277–3305. http://ieeexplore.ieee.org/stamp/stamp.jsp?tp=&arnumber=7867830&isnumber=7905784
27. Branciard C (2016) Witnesses of causal nonseparability: an introduction and a few case studies. Sci Rep 6:26018
28. Pearl J (2009) Causality: models, reasoning and inference. Cambridge University Press
29. Hardy L (2009) Operational structures as a foundation for probabilistic theories, PIRSA:09060015, Talk at http://pirsa.org/09060015/
30. Chiribella G, D'Ariano GM, Perinotti P (2010) Probabilistic theories with purification. Phys Rev A 81:062348
31. Chiribella G, D'Ariano GM, Perinotti P (2011) Informational derivation of quantum theory. Phys Rev A 84:012311
32. Hardy L (2011) Reformulating and reconstructing quantum theory. arXiv:1104.2066
33. Jamiołkowski A (1972) Linear transformations which preserve trace and positive semidefiniteness of operators. Rep Math Phys 3(4):275–278
34. Choi M-D (1975) Completely positive linear maps on complex matrices. Lin Alg Appl 10:285–290
35. Barnum H, Fuchs CA, Renes JM, Wilce A (2005) Influence-free states on compound quantum systems. arXiv:quant-ph/0507108
36. Eddington A (1928) The nature of the physical world. Cambridge University Press
37. Davies E, Lewis J (1970) An operational approach to quantum probability. Comm Math Phys 17:239–260
38. Kraus K (1983) States, effects and operations. Springer, Berlin
39. Chiribella G, D'Ariano GM, Perinotti P (2009) Theoretical framework for quantum networks. Phys Rev A 80:022339
40. Bennett CH, Brassard G, Crépeau C, Jozsa R, Peres A, Wootters WK (1993) Teleporting an unknown quantum state via dual classical and Einstein-Podolsky-Rosen channels. Phys Rev Lett 70:1895
41. Feix A, Araújo M, Brukner Č (2016) Causally nonseparable processes admitting a causal model. New J Phys 18: 083040. http://iopscience.iop.org/1367-2630/18/8/083040/
42. Bell JS (1964) On the Einstein Podolsky Rosen Paradox. Physics 1(3):195–200
43. Werner RF (1989) Quantum states with Einstein-Podolsky-Rosen correlations admitting a hidden-variable model. Phys Rev A 40:4277
44. Knill E (1996) Conventions for quantum pseudocode, LANL report LAUR-96-2724
45. Valiron B, Selinger P (2005) A lambda calculus for quantum computation with classical control. In: Proceedings of the 7th international conference on typed lambda calculi and applications (TLCA), vol 3461. Lecture Notes in Computer Science, pp. 354–368. Springer
46. Chiribella G, D'Ariano GM, Perinotti P (2008) Transforming quantum operations: quantum supermaps. Europhys Lett 83:30004
47. Ried K, Agnew M, Vermeyden L, Janzing D, Spekkens RW, Resch KJ (2015) A quantum advantage for inferring causal structure. Nat Phys 11:414–420
48. Araújo M, Feix A, Navascués M, Brukner Č (2017) A purification postulate for quantum mechanics with indefinite causal order. Quantum 1:10

Chapter 3
Witnessing Causal Nonseparability: Theory and Experiment

Theory Section

3.1 Back Story

This first part of the chapter is based on Ref. [1]. At the time, we knew very well
the process matrix formalism as presented in the original paper [2]. The observation
that the set of causally separable process matrices is convex, was the start of this
work. This is important, because of the well-known hyperplane separation theorem
that is greatly used for entanglement witnesses: since the state of separable states
define a convex set, then, by the hyperplane separation theorem, there is a hyperplane
separating any non-separable state from the convex set. The hyperplane corresponds
to measurements on the non-separable state, to witness its nonseparability. Using this
theorem in the space of process matrices means that there is a hyperplane separating
a causally non-separable process matrix from the convex set of causally separable
ones. The hyperplane in this case corresponds to operations to be performed by the
parties in order to *witness causal nonseparability*.

But why would we pursue this project, as it seems to be simply a mathematical theorem applied in the process matrix framework, much like entanglement witnesses is
the result of the application of the framework to quantum mechanics? The answer lies
in the experimental realizations. Entanglement witnesses are extremely useful whenever entanglement needs to be certified. Using this simple mathematical theorem, an
experimenter instead of having to perform informationally complete measurements
for a full-state tomography to verify entanglement, an entanglement witness requires
a smaller amount of measurements. Similarly, in an experimental realization of a
process matrix, when the interest is whether it is separable or not, full process matrix
tomography requires informationally complete operations of all parties; whereas
causal separability witnesses restrict the amount of the parties operations. Hence, we
had the tools and the motivation to pursue this project.

© Springer Nature Switzerland AG 2019
C. Giarmatzi, *Rethinking Causality in Quantum Mechanics*, Springer Theses,
https://doi.org/10.1007/978-3-030-31930-4_3

3.2   Introduction

As we have seen in Chap. 2, using the process matrix framework, it is in principle possible to imagine situations where the causal order of quantum operations is
not fixed in advance, whether it is probabilistic, dynamical or even indefinite [2,
3]. The motivation to study the exotic situations of indefinite causal order comes
from a foundational point of view for general probabilistic theories [4–7], but also
shown recently to pave the way for novel quantum architectures with computational
advantages [8]. The main example of the latter is the technique called "quantum
switch", analyzed in Chap. 2, in Sect. 2.4 (as a causally nonseparable process matrix)
whose computational advantage has been proved [9, 10] and whose experimental
implementation has been realized [11, 12].

This chapter (the theory section) is based on the publication in Ref. [1] and is about
applying the hyperplane separation theorem to process matrices, to create witnesses
of causal non separability. In other words, we develop a device-dependent (we will
see later what that means) way to detect causal nonseparability of a process matrix.
Before we go into technicalities, let us present the basic ideas. Think of a circle drawn
in some (x, y) coordinate space. Then it is easy to say that for every point (x', y')
that is not inside or in the boundaries of this circle, there must be a straight line that
separates the circle from this point. The same idea can be applied to a sphere with
the difference that now the line would be replaced by a two-dimensional plane. For a
d-dimensional sphere there will be a hyperplane separating the sphere from any outer
point. Finally, we can even think this further and possibly make a claim that the same
idea applies to any convex set (instead of a sphere) at an arbitrary dimensional space:
for any point outside a convex set there is a hyperplane separating this point from
the set. Fortunately this claim has been already proven and it is called 'hyperplane
separation theorem' [13].

This beautiful theorem, depicted in Fig. 3.1 is applied to the space of quantum
states: since the set of separable states form a convex set, for every nonseparable
state there exists a hyperplane separating the set from the given state. Translating
this geometrical result into equations, we get that for every nonseparable state ρ_{nsep},
there exist an operator S such that $\text{Tr}(S\rho_{\text{nsep}}) < 0$ while $\text{Tr}(S\rho_{\text{sep}}) \geq 0$ (the zero is
arbitrary). The operator S is the geometrical description of the separating hyperplane.
Now remember the Born rule: the probability of an outcome i out of a set of possible
outcomes for a given quantum measurement described by the POVM elements $\{E_i\}$
is given by $\text{Tr}(E_i\rho)$. Hence, we can see the hyperplane S as some operator which
we apply to our state and it can correspond to a single or a mixture of measurement
operators. Now we see that the theorem states that for every nonseparable state there
exists an operator that we can apply to the state and if the result is smaller than 0
that proves its nonseparability. From an experimental perspective this is great news;
it states that quantum state tomography is no longer needed to verify nonseparability
and this is particularly useful in the case of entanglement which is usually what we
need to verify—in this case the object S is called *entanglement witness*. Moving to

Fig. 3.1 Sketch of the hyperplane separation theorem: for every point outside the convex set, there must be a hyperplane separating the point from the set

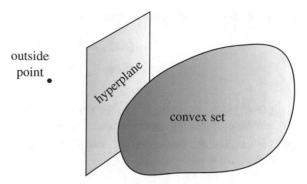

the space of process matrices, since the set of separable process matrices defines a convex set, we can define later, by analogy, a *causal witness*.

As any member of the police can tell you, for a witness to exist it is one thing; for a witness to be found it is a different one. Entanglement witnesses are hard to find. The good news is that causal witnesses (in the cases we studied) can always be found efficiently. We studied the general bipartite case and a particular tripartite case. This is because of the peculiarity of the general tripartite and multipartite case studied in Chap. 2, based on Ref. [3], which makes the problem more intricate. Back to the good news, the problem of finding a causal witness can be formulated as a SemiDefinite Program (SDP) which, using convex optimization techniques, can be solved efficiently. This means that, upon an experimental implementation of a causally nonseparable process matrix, (if it lies within our studied cases) we can find a causal witness (the set of operations the parties have to perform) to prove its causal nonseparability. This is particularly interesting in light of the two recent experimental realizations of a tripartite causally nonseparable process matrix in Refs. [11, 12] and in our own implementation which will be discussed later in this chapter.

3.3 Mathematical Characterization of Valid and Causally Separable Processes

Let us review briefly the process matrix formalism, and move to the characterization of valid bipartite and tripartite process matrices in terms of linear constraints on the process matrix. We are considering scenarios where experimenters (parties) are located inside closed laboratories A, B, ... and perform experiments on incoming systems A_I, B_I, ..., obtain outcomes i, j, ... and send out outgoing systems A_O, B_O, ... assigned to Hilbert spaces $\mathcal{H}^X$, of dimension d_X, with X being the corresponding system. No assumption is made on the causal order of the laboratories. The operation of each lab is described by a Completely Positive (CP) map from the input to the output which yields the outcome i, $\mathcal{M}_i^A$, isomorphic to its CJ matrix $M_i^{A_I A_O}$. The joint probability for a specific set of local CP maps to be realized by the parties $A, B \cdots$ is then given by the following linear function

$$p(\mathcal{M}_i^A, \mathcal{M}_j^B, \cdots | \{\mathcal{M}_i^A\}, \{\mathcal{M}_j^B\}, \cdots)$$
$$= \mathrm{Tr}\left[W^{A_I A_O B_I B_O \cdots}\left(M_i^{A_I A_O} \otimes M_j^{B_I B_O} \otimes \cdots\right)\right], \tag{3.1}$$

with $\{\mathcal{M}_i^X\}$ being the collection of the possible CP maps for party X which sum up to a CPTP map, $\mathcal{M}^X = \sum_j \mathcal{M}_j^X$ representing a quantum instrument, whose CJ matrix $M^{X_I X_O}$ satisfies the normalization condition $\mathrm{Tr}_{X_O} M^{X_I X_O} = \mathbb{1}^{X_I}$. For conveniency, we define two maps:

$$_x W = \tfrac{\mathbb{1}^x}{d_x} \mathrm{Tr}_x W \tag{3.2}$$
$$_{[1-x]} W = W - {}_x W, \tag{3.3}$$

where x is any subsystem in W. Then the condition for the CJ matrix $M^{X_I X_O}$ to represent a trace preserving map is written as $_{X_O} M^{X_I X_O} = \mathbb{1}^{X_I X_O}/d_{X_O}$.

Bipartite case: In the Appendix A, we provide the characterization of the set of valid bipartite process matrices. The result is that an operator $W \in A_I \otimes A_O \otimes B_I \otimes B_O$ is valid if and only if $W \geq 0$, $\mathrm{Tr}\, W = d_{A_O} d_{B_O}$ and $W \in \mathcal{L}_{V_2}$, where $\mathcal{L}$ is the linear subspace defined by the projector

$$L_V(W) = {}_{A_O} W + {}_{B_O} W - {}_{A_O B_O} W - {}_{B_I B_O} W + {}_{A_O B_I B_O} W - {}_{A_I A_O} W + {}_{A_I A_O B_O} W. \tag{3.4}$$

In terms of simple individual conditions on the process matrix, we have that a valid process matrix must satisfy

$$W \geq 0$$
$$\mathrm{Tr}\, W = d_{A_O} d_{B_O}$$
$$_{B_I B_O} W = {}_{A_O B_I B_O} W \tag{3.5}$$
$$_{A_I A_O} W = {}_{A_I A_O B_O} W$$
$$W = {}_{B_O} W + {}_{A_O} W - {}_{A_O B_O} W.$$

Now let us move to the characterization of the causally separable process matrices. As we have seen in Chap. 2 and in Ref. [2] they have the form

$$W_{csep} = q W^{A \prec B} + (1 - q) W^{B \prec A}, \tag{3.6}$$

with $q \in [0, 1]$, where each term is a valid process matrix compatible with one party being in the causal past of the other and both compatible with the parties being causally independent ($A||B$). These terms must satisfy some constraints, as we have already seen in Chap. 2, namely they have identity on the output system of the last party. Using the notation introduced in Eq. (3.3), these constraints can be expressed as

$$W^{A \prec B} = {}_{B_O}W^{A \prec B}$$
$$W^{B \prec A} = {}_{A_O}W^{B \prec A}. \tag{3.7}$$

Note that $W^{A||B}$ will satisfy both constraints. Hence, ignoring for conveniency the normalization constraint, a valid process matrix is causally separable if and only if it admits the decomposition (3.6) where each term satisfies the corresponding constraint (3.7). Hence, the set of causally separable process matrices is a convex cone, which we denote as $\mathcal{W}_{csep}$. A process matrix that cannot be decomposed as in (3.6) is called causally nonseparable.

Tripartite case: In the case of three parties, the conditions that define the subspace of valid process matrices is obtained in a similar fashion.

$$\begin{aligned} {}_{X_O}W &= 0 \\ {}_{X_O Y_O}W &= 0 \\ {}_{X_O Y_O Z_O}W &= 0 \\ {}_{[1-X_O]}\text{Tr}_{Y_I Y_O}W &= 0 \\ {}_{[1-X_O]}\text{Tr}_{Y_I Y_O Z_I Z_O}W &= 0 \end{aligned} \tag{3.8}$$

with $X, Y, Z = A, B, C$, which define a linear subspace denoted as $\mathcal{L}_V$. Hence, an operator $W \in A_I \otimes A_O \otimes B_I \otimes B_O \otimes C_I \otimes C_O$ is valid if and only if $W \geq 0$, $\text{Tr}\, W = d_{A_O} d_{B_O} d_{C_O}$ and $W = L_V(W)$ as defined in the Appendix.

Let us now move to the tripartite causally separable process matrices, but with the extra condition that the output system of one party, say C, is trivial, i.e. $d_{C_O} = 1$. This means that the party cannot signal to any other, and can therefore be considered as last. This leaves us with two possible causal configurations, i.e. $A \prec B \prec C$ and $B \prec A \prec C$. A tripartite causally separable process matrix with the constraint that Charlie's output is trivial, can then be written as a convex combination of process matrices compatible with the two causal configurations.

$$W_{csep}^{3C} = q W^{A \prec B \prec C} + (1-q) W^{B \prec A \prec C}, \tag{3.9}$$

with $q \in [0, 1]$. Ignoring again the normalization constraint, this defines a convex cone, which we denote as $\mathcal{W}_{csep}^{3C}$, with the $3C$ indicating that we are in the special case where Charlie's output is trivial (or last).

A process matrix compatible with each causal configuration should satisfy the following three constraints. The intuition behind them is that, for example, for $A \prec B \prec C$, (a) there is identity on the output of C, (b) upon tracing out C, we get identity on the output of B, and (c) tracing out B, C we get identity on the output of A. (These constraints are equivalent to defining a 'quantum comb' [14, 15], an object similar to the process matrix for a fixed causal order between the parties.)

$$[1-C_O] W^{A \prec B \prec C} = 0$$
$$[1-B_O] \operatorname{Tr}_{C_I C_O} W^{A \prec B \prec C} = 0$$
$$[1-A_O] \operatorname{Tr}_{B_I B_O C_I C_O} W^{A \prec B \prec C} = 0$$
$$[1-C_O] W^{B \prec A \prec C} = 0 \qquad (3.10)$$
$$[1-A_O] \operatorname{Tr}_{C_I C_O} W^{B \prec A \prec C} = 0$$
$$[1-B_O] \operatorname{Tr}_{A_I A_O C_I C_O} W^{B \prec A \prec C} = 0.$$

These three conditions for each process matrix, define a linear subspace with projectors $L_{A \prec B \prec C}$ and $L_{B \prec A \prec C}$ such that

$$W^{A \prec B \prec C} = L_{A \prec B \prec C}(W^{A \prec B \prec C}) \quad \text{and} \quad W^{B \prec A \prec C} = L_{B \prec A \prec C}(W^{B \prec A \prec C}) \quad (3.11)$$

The set of tripartite causally separable process matrices is a closed convex set, which we denote as $\mathcal{W}_{csep}^{3C}$.

Now that we have our formal characterizations of bipartite causally separable process matrices ($\mathcal{W}_{csep}$) and tripartite ones with one party having a trivial output, ($\mathcal{W}_{csep}^{3C}$), we are ready to formulate causal witnesses for these cases. Namely, we will be able to detect causal nonseparability in the general bipartite case and in the tripartite case where Charlie's output is trivial.

3.4 General Bipartite Causal Witnesses

We begin with the definition: we call a Hermitian operator S a causal witness (or witness for brevity) if

$$\operatorname{Tr}(S W_{csep}) \geq 0 \qquad (3.12)$$

for every causally separable process matrix W_{csep}. We remind the hyperplane separation theorem [13] applied in our case: the set of causally separable process matrices is closed and convex; therefore for every causally nonseparable process matrix there exists a witness (defined just above) such that $\operatorname{Tr}(S W_{nsep}) \leq 0$. Since this operator actually *witnesses* the nonseparability of W_{nsep} we can call it a Witness for this W_{nsep}, to make it distinct from the set of all witnesses (see Fig. 3.2). We proceed to the mathematical characterization of the set of causal witnesses, in terms of linear constraints they should satisfy. This will allow us to formulate our problem as an SDP. In the bipartite case we remind that $W_{csep} = q W^{A \prec B} + (1 - q) W^{B \prec A}$. Then the above equation is equivalent to

$$\operatorname{Tr}(S W^{A \prec B}) \geq 0, \quad \forall W^{A \prec B} \qquad (3.13a)$$
$$\operatorname{Tr}(S W^{B \prec A}) \geq 0, \quad \forall W^{B \prec A}. \qquad (3.13b)$$

**Fig. 3.2 The set of causally
separable process matrices
and a witness**: for every
causally nonseparable
process matrix, there must be
a witness separating its point
from the set of causally
separable process matrices

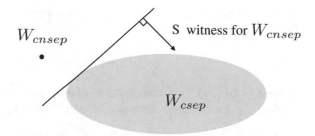

Let us focus on Eq. (3.13a). Remember the first of the Eq. (3.7), $W^{A \prec B} =_{B_O} W^{A \prec B}$, meaning that a valid bipartite process matrix W with identity on the output of B is compatible with a causal configuration $A \prec B$. Then we can write

$$\text{Tr}(S_{B_O} W) \geq 0 \quad \forall W \in \mathcal{L}_V, \ W \geq 0. \tag{3.14}$$

Remark that the map $_X W := \mathbb{1}^X \text{Tr}_X W$ is self-adjoint, which means that $\langle _X S, W \rangle = \langle S, _X W \rangle$, and considering the trace as a Hilbert-Schmidt inner product, $\langle S, W \rangle = \text{Tr}(S, W^*)$, we have that

$$\text{Tr}(S_{B_O} W) = \text{Tr}(_{B_O} S W). \tag{3.15}$$

For the right-hand side to be non-negative for all valid process matrices, it is sufficient that $_{B_O} S \geq 0$, and similarly $_{A_O} S \geq 0$ is sufficient for the Eq. (3.13b) to hold. Hence, for S to be a witness it is sufficient that

$$_{B_O} S \geq 0, \ \text{and} \ _{A_O} S \geq 0. \tag{3.16}$$

Note that adding an operator $S^\perp$ that belongs to the orthogonal complement $\mathcal{L}_V^\perp$ of $\mathcal{L}_V$ (remember $W \in \mathcal{L}_V$) to any witness gives another valid witness, since $\text{Tr}[(S + S^\perp)W] = \text{Tr}(SW)$ for any valid $W \in \mathcal{L}_V$. It turns out that this suffices to completely characterize the set of causal witnesses, as we state in the following theorem.

Theorem *A Hermitian operator $S \in A_I \otimes A_O \otimes B_I \otimes B_O$ is a (causal) witness if and only if S can be written as*

$$S = S_P + S^\perp \tag{3.17}$$

where each term is a Hermitian operator such that

$$_{B_O} S_P \geq 0, \quad _{A_O} S_P \geq 0, \quad L_V(S^\perp) = 0 \tag{3.18}$$

The proof of this theorem is in the Appendix C of the relevant paper [1]. This theorem provides a characterization of the set of causal witnesses, which is a closed convex cone which we denote as $\mathcal{S}$.

Note that the operator $S^\perp$ can be arbitrarily chosen as it does not change the value $\mathrm{Tr}(SW)$, for instance it can be

$$S^\perp = L_V(S_P) - S_P \tag{3.19}$$

so that $S = L_V(S_P)$. This restricts the set of witnesses to the subspace of valid process matrices $\mathcal{L}_V$. Hence, these witnesses have the following characterization.

Corollary 4 *A Hermitian operator $S \in \mathcal{L}_V$ is a witness if and only if there exists a Hermitian operator $S_P \in A_I \otimes A_O \otimes B_I \otimes B_O$ such that $S = L_V(S_P)$, $_{B_O} S_P \geq 0$, and $_{A_O} S_P \geq 0$.*

This set of causal witnesses is also a closed convex cone, and we denote as $\mathcal{S}_V = \mathcal{S} \cap \mathcal{L}_V$.

Now that we have the characterization of the set of causal witnesses in the general bipartite case we are ready to formulate the problem of checking causal nonseparability of a W and finding a Witness for it, as an SDP [16].

3.5 Tripartite Causal Witness for a Special Case

In a previous section, we characterized the set of causally separable tripartite process matrices, for the particular case where one party is last (has a trivial output system). For these cases, we can find a witness, i.e. a Hermitian operator S, such that

$$\mathrm{Tr}(SW_{csep}^{3C}) \geq 0 \tag{3.20}$$

for every causally separable process matrix W_{csep}^{3C}, with the $3C$ indicating the special case where one party has a trivial output system (a condition that holds for the whole section). Since $\mathcal{W}_{csep}^{3C}$ is a closed convex set, there must be a witness such that $\mathrm{Tr}(SW_{nsep}^{3C}) < 0$, for every causally nonseparable W_{nsep}^{3C}.

From Eq. (3.9) we have the explicit decomposition $W_{csep}^{3C} = qW^{A \prec B \prec C} + (1 - q)W^{B \prec A \prec C}$ with each term satisfying the set of constraints of Eq. (3.10) or, in terms of projectors, $W^{A \prec B \prec C} = L_{A \prec B \prec C}(W^{A \prec B \prec C})$ and $W^{B \prec A \prec C} = L_{B \prec A \prec C}(W^{B \prec A \prec C})$. Similarly to the bipartite case, the characterization of the set of causal witnesses, denoted as $\mathcal{S}^{3C}$ is the following

Theorem *A Hermitian operator $S \in A_I \otimes A_O \otimes B_I \otimes B_O \otimes C_I \otimes C_O, d_{C_O} = 1$ is a causal witness if and only if S can be written as*

$$\begin{aligned} S &= S_{ABC}^P + S_{ABC}^\perp = S_{BAC}^P + S_{ABC}^\perp, \\ S_{ABC}^P &\geq 0, \quad L_{A \prec B \prec C}(S_{ABC}^\perp) = 0 \\ S_{BAC}^P &\geq 0, \quad L_{B \prec A \prec C}(S_{BAC}^\perp) = 0 \end{aligned} \tag{3.21}$$

with $L_{A \prec B \prec C}$ and $L_{B \prec A \prec C}$ being the projectors onto subspaces defined in Eq. (3.10). These restrictions define the set of causal witnesses for this special tripartite case, which we denote as $\mathcal{S}^{3C}$.

The proof of this theorem is given in the Appendix G of the relevant paper [1].

3.6 SDP for Witnessing Causal Nonseparability

We are now ready to formulate the SDP; given the complete characterization of the convex cone of the causal witnesses, and that of the valid and causally separable process matrices, it will be able to identify if a given W is causally separable or not (for the cases we studied above: the general bipartite and one particular tripartite). If it is, the SDP will provide the explicit decomposition in causally ordered process matrices. If the input W is causally nonseparable, then it provides the causal Witness for it.

There are two ways to formulate our problem which we call 'primal' and 'dual' SDP. The primal SPD is asking the following: how much *noise* can I add to the (possibly nonseparable) process matrix W before it becomes causally separable? There are different ways to represent this noise which in a practical implementation would depend on the experimental components used to create the process matrix (gates, channels, etc.). Here we will consider the case of 'white noise', represented by a process matrix $\mathbb{1}° = \mathbb{1}/d_I$, where d_I is the product $d_{A_I} d_{B_I} \cdots$. This white-noise process matrix represents a situation where all parties receive a joint input system in a maximally mixed state. Therefore, for a given W, we consider the 'noisy' process matrix

$$W(r) = \frac{1}{1+r}(W + r\mathbb{1}°) \tag{3.22}$$

and ask the value of r such that it is causally separable (see Fig. 3.3). Ignoring the normalization of $W(r)$—remember that the cone of causally separable process matrices $\mathcal{W}_{csep}$ was defined without the normalization constraint—we define the following optimization problem

$$\begin{aligned} &\texttt{minimize} \quad r \\ &\texttt{such that} \quad W + r\mathbb{1}° \in \mathcal{W}_{csep} \end{aligned} \tag{3.23}$$

As we have seen, the characterization of the cone $\mathcal{W}_{csep}$ yields linear constraints on the operator $W + r\mathbb{1}°$. Therefore, this optimization problem defines an SDP problem. The solution to this problem is the optimal (minimum) value of $r = r^*$ and an explicit decomposition $W(r^*)$ as a convex combination $W(r^*) = W^{A \prec B} + W^{B \prec A}$, of non-normalized process matrices compatible with one of the two possible causal orders. Any positive value of r^* would mean that the input W is nonseparable. In other words, if we need to add any amount of $\mathbb{1}°$ to make the input process matrix

Fig. 3.3 Optimal witness with respect to 'white' noise: The value r^* is the amount of white noise $\mathbb{1}^\circ$ that needs to be added to W_{cnsep} to make it causally separable

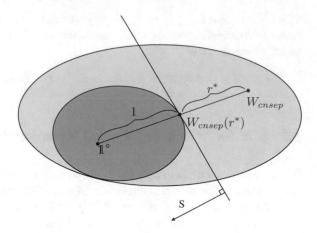

causally separable, then it means that it is causally nonseparable. Equation (3.23) is the formulation of the 'primal' problem.

We now present the 'dual' (to the 'primal') problem.

$$\texttt{minimize} \quad \mathrm{Tr}(SW)$$
$$\texttt{such that} \quad S \in \mathcal{S}_V \quad \text{and} \quad \mathrm{Tr}(S\mathbb{1}^\circ) = 1 \tag{3.24}$$

We ask to minimize the value $\mathrm{Tr}(SW)$, such that the operator S is a causal witness (belonging to the closed convex cone of causal witnesses $\mathcal{S}_V = \mathcal{S} \cap \mathcal{L}_V$). We restrict our search to $\mathcal{S}_V$, instead of $\mathcal{S}$, because sometimes it may be the case that in practice it is more convenient: optimizing over the larger space $\mathcal{S}$ can make some solvers unstable. However, if one wishes to make such a search, the result of the above SDP would be equivalent to both spaces as the values $\mathrm{Tr}(SW)$ and $\mathrm{Tr}(S\mathbb{1}^\circ)$ would remain the same by adding a term $S^\perp \in \mathcal{L}_V^\perp$ to the witness S. A constraint on $\mathrm{Tr}(S\mathbb{1}^\circ)$ is needed to avoid obtaining an optimal value of $-\infty$ for the minimized value $\mathrm{Tr}(SW)$. We chose $\mathrm{Tr}(S\mathbb{1}^\circ) = 1$ because this makes the output of the dual SDP to be a meaningful value, as we will see just below.

The optimal solution of the above dual problem is S^* and it is a causal Witness if the minimized value of $\mathrm{Tr}(SW)$ is negative. From the Duality Theorem for SDP problems [16], we have that the solutions of the primal and dual problem are related through the expression

$$r^* = -\mathrm{Tr}(S^*W). \tag{3.25}$$

To understand the value of this relationship, let us imagine that the input process matrix W is causally nonseparable. Now recall that r^* is the minimum amount of $\mathbb{1}^\circ$ that needs to be added to the input W to make it causally separable, and that $\mathrm{Tr}(S\mathbb{1}^\circ) = 1$. Then we have that

$$\text{Tr}(S^*W(r)) = \frac{\text{Tr}(S^*W)}{1+r} + \frac{r\,\text{Tr}(S^*\mathbb{1}^\circ)}{1+r} = -\frac{r^*}{1+r} + \frac{r}{1+r} < 0 \quad \forall r < r^*.$$

(3.26)

This means that for all $r < r^*$ (the process matrix $W(r)$ is causally nonseparable), $\text{Tr}(S^*W(r))$ would still be negative and thus S^* would still detect the causal nonseparability of $W(r)$ and would still be a Witness for it. This means that this Witness is optimal for $W(r)$, that is for W that is subjected to white noise, see Fig. 3.2.

Note that we can define other noise models, other that the above 'white noise' model. For example we can replace the above process $\mathbb{1}^\circ$ with another process matrix W° that for some reason the experimentalist is competing against, when trying to implement the desired process matrix W. Then in the above dual problem the normalization constraint would be $\text{Tr}(SW^\circ) = 1$ and one can show, as in Ref. [1], that as long as W° is in the interior of $\mathcal{W}_{csep}$ the SDP would still be solved efficiently, and that the optimal solutions of the primal and dual problem would still satisfy Eq. (3.25).

Due to the fact that the SDPs are providing the optimal witness, in the general case of some noise model, this value $r*$ or $-\text{Tr}(S^*W)$ gives us a 'measure' of the nonseparability of the process matrix W. We define the *random robustness* of W as its resistance to white noise $\mathbb{1}^\circ$

$$R_r(W) = -\text{Tr}(S^*W),$$

(3.27)

where the optimal witness is the one obtained with respect to white noise, that is, from an SDP as written above. In the case where the white noise model is the right one for some particular implementation of the given W, the random robustness is a good measure of the causal nonseparability of W. However, one has to be careful because if the noise model does not fully capture the experimental imperfections, then it is not guaranteed that the random robustness is a good measure.

3.7 Implementing a Causal Witness in the Lab

In this section we are interested in practical scenarios, where we have implemented in the lab a causally nonseparable process matrix for a number of parties, and we wish to prove its causal nonseparability through a causal Witness. We will see in the next chapter that this is exactly our goal for a specific experimental implementation for a particular causally nonseparable process matrix, but let us first lay down the basics for such a task.

First of all, what does it mean to 'measure' a Witness? From the definition of the Witness we know that, for a given causally nonseparable W_{nsep}, we have that $\text{Tr}(SW_{nsep}) < 0$ whereas $\text{Tr}(SW_{csep}) \geq 0$ for all causally separable W_{csep}. But how can we verify experimentally that the process matrix W_{nsep} that (we think) we have implemented in the lab is indeed causally nonseparable?

Let us remind some basics of the process matrix formalism. The scenario is: we have a set of parties performing operations, whose outcomes are correlated through their environment, represented by the process matrix. Each operation, say that of party A, is described by a Completely Positive (CP) map, that maps their input A_I to their output A_O system, yielding an outcome i, $\mathcal{M}_i^A$, represented by their Choi-Jamiołkowski (CJ) operator $M_i^{A_I A_O}$, whose collection represents the settings of each party. We have that the joint probability for a set of parties to realize these maps, given their settings, is given by

$$p(M_i^{A_I A_O}, M_j^{B_I B_O}, \cdots |\{M^{A_I A_O}\}, \{M^{B_I B_O}\}, \cdots) = \mathrm{Tr}[(M_i^{A_I A_O} \otimes M_j^{B_I B_O} \otimes \cdots) W^{A_I A_O B_I B_O \cdots}]$$

(3.28)

Back to our task of measuring our Witness, we see that since we have to prove that for a given W_{nsep}, $\mathrm{Tr}(S W_{nsep}) < 0$ all we have to do, is to find a way to express the witness as

$$S = \sum_{i,j\cdots} \alpha_{i,j,\cdots} (M_i^{A_I A_O} \otimes M_j^{B_I B_O} \otimes \cdots).$$

(3.29)

Each term in this sum corresponds to the experimentally accessible probabilities

$$p(M_i^{A_I A_O}, M_j^{B_I B_O} \cdots |\{M^{A_I A_O}\}, \{M^{B_I B_O}\}, \cdots)$$

(3.30)

and it will be such that

$$\sum_{i,j,\cdots} \alpha_{i,j,\cdots} p(M_i^{A_I A_O}, M_j^{B_I B_O} \cdots |\{M^{A_I A_O}\}, \{M^{B_I B_O}\}, \cdots)) = \mathrm{Tr}(S W_{nsep}) < 0.$$

(3.31)

Therefore, experimentally measuring the operator S and proving its sign is negative, proves the causal nonseparability of W_{nsep}.

So now our task is to decompose the witness as a weighted sum of experimentally accessible operations for the parties, which would of course vary in each experimental implementation of the W_{nsep}. For instance, it may be the case that some of the parties can only perform unitary operations, as opposed to measure-and-prepare operations. In that case, we can insert such a restriction to the SDP and ask to find a witness such that it can be decomposed as a weighted sum of unitary operations for the parties. In particular, we modify the 'dual' problem (3.24) and replace $\mathcal{S}_V$ by the set $\mathcal{S}_U \subset \mathcal{S}$, where $\mathcal{S}_U$ denotes the space of causal witnesses, for which some of the parties perform unitary operations. The price to pay is that by restricting to this subspace we may not obtain the optimal witness, or even a witness at all, because such a witness will not be able to witness all causally nonseparable process matrices. However, restricting our search for a witness in such a way is a powerful tool as we shall see in the next sections.

3.8 Witness for the Quantum Switch

The quantum switch, as we have seen in Chap. 2, is a technique which makes the order in which two quantum gates are realized to be conditioned on the state of a quantum bit [8]. In particular, imagine that two parties (that perform the quantum gates), A and B, each receive an input system which we call the target bit, perform unitary operations, U_A, U_B respectively, and send the system out. The order in which the parties perform their unitaries is conditioned on a control bit: if its state is $|0\rangle$, the target bit undergoes the unitary U_A before the unitary U_B, if the state is $|1\rangle$, the order of the unitaries is reversed (later there will be a nice Figure for this). After the operations of these two parties, whatever their order were, there is a third party C that receives the control and target bit.

Seen through the process matrix formalism, the circuit of the quantum switch is a tripartite causally nonseparable process matrix. Interestingly, the original idea was developed as a novel quantum circuit architecture that goes beyond the standard quantum circuit model [8]—indeed the quantum switch cannot be represented as standard quantum circuit where the order of operations occur in well-defined time-slots. However, when it was later re-examined in Refs. [1, 3], it was the only example of a tripartite causally nonseparable process matrix that has a direct physical implementation, as shown in Refs. [11, 12] and as we will see in the next section.

Now that we have a nice tripartite causally nonseparable process matrix with a physical implementation, we can search for a Witness for it. Recall that in order to write an SDP to witness the causal nonseparability of an $n-$partite process matrix, we need the characterization of the set of $n-$partite causally separable process matrices. This is why, in a section above, we characterized the cone of tripartite causally separable process matrices in the special case where one party is always last (having a trivial output).

3.8.1 Optimal Witness Through SDP

As seen in Chap. 2, Sect. 2.4 the process matrix of the quantum switch is

$$W_s^{A_I A_O B_I B_O C_I} = |W\rangle\langle W|^{A_I A_O B_I B_O C_I}, \tag{3.32}$$

where

$$|W\rangle^{A_I A_O B_I B_O C_I} = (|0\rangle^{C_I^c}|\psi\rangle^{A_I}|\Phi^+\rangle^{A_O B_I}|\Phi^+\rangle^{B_O C_I^s} + |1\rangle^{C_I^c}|\psi\rangle^{B_I}|\Phi^+\rangle^{B_O A_I}|\Phi^+\rangle^{A_O C_I^s})/\sqrt{2}, \tag{3.33}$$

where C_I^c is the control bit that is input to C, and C_I^s is the target bit or system, that C receives after the operations of A and B. Here, we will consider the above process matrix, but with the system C_I^s traced out. This is because the remaining process matrix is still causally nonseparable and it reduces its dimensions, which becomes crucial to the computational solvers of the SDP. Note that the output system of Charlie has dimension 1. Note also that if we trace out the only system of C that we consider, C_I^c, the remaining process matrix becomes causally separable—it is that extra party in the future of A and B that makes the process matrix causally nonseparable. Therefore, the process matrix we end up with, is

$$W_{switch} = \text{Tr}_{C_I^s}(W_s^{A_I A_O B_I B_O C_I}). \tag{3.34}$$

Then the primal SDP, as formulated in (3.23), is written as

$$\begin{aligned} &\texttt{minimize} \quad r \\ &\texttt{such that} \quad W_{switch} + r\mathbb{1}^\circ \in \mathcal{W}_{csep}^{3C}, \end{aligned} \tag{3.35}$$

where $\mathcal{W}_{csep}^{3C}$ is the cone of causally separable tripartite process matrices with the condition that one party's output is trivial, as defined by Eqs. (3.9) and (3.10).

The dual problem is written as

$$\begin{aligned} &\texttt{minimize} \quad \text{Tr}(SW_{switch}) \\ &\texttt{such that} \quad S \in \mathcal{S}^{3C} \quad \text{and} \quad \text{Tr}(S\mathbb{1}^\circ) = 1 \end{aligned} \tag{3.36}$$

where $\mathcal{S}^{3C}$ is the set of causal witnesses defined for this case of three parties in Eq. (3.21). Here we could use $\mathcal{S}_V^{3C} = \mathcal{S}^{3C} \cap \mathcal{L}_V$ but it turns out that both cases are solved quite fast. Also, although $\mathcal{S}^*$ and $\mathcal{S}_V^{3C}$ give a different witness, its general robustness and visibility remain the same.

We solved the primal and dual problem and obtained the random robustness of the quantum switch (its resistance to white noise)

$$R_r(W_{switch}) = r_{switch}^* \approx -1.57603 \tag{3.37}$$

Another way to estimate this result is in terms of the 'visibility' v which is related to the value r as in

$$v = \frac{1}{1+r} \tag{3.38}$$

Then the noisy quantum switch

$$W_{switch}(v) = vW_{switch} + (1-v)\mathbb{1}^\circ, \tag{3.39}$$

is causally nonseparable for all $v > v_{switch}^* = \frac{1}{1+r_{switch}^*} \approx 0.3882$.

Curious unpublished result: As we have discussed, we only considered a special tripartite case: one party has a trivial output, or in other words, one party is last, say C. We did so because in this case we can write the form of a causally separable process matrix in a simple form: it is a mixture of process matrices compatible with one of the causal configurations $A \prec B \prec C$ or $B \prec A \prec C$. Another reason why we restricted to this case is because when we search for a witness for the quantum switch, we can make that assumption since the quantum switch does have Charlie always last.

However, as we have seen in Chap. 2, the general form of a tripartite (extensibly, we comment on that later) causally separable process matrix is the following

$$W_{ecsep} = q_1 W_{ecsep;(A,B)\not\prec C} + q_2 W_{ecsep;(A,C)\not\prec B} + q_3 W_{ecsep;(B,C)\not\prec A},$$

$$q_i \geq 0, \ \forall i = 1, 2, 3, \ \sum_{i=1}^{3} q_i = 1, \tag{3.40}$$

where $W_{ecsep;(A,B)\not\prec C}$ is a valid process matrix compatible with $(A, B) \not\prec C$ and has the form

$$W_{ecsep;(A,B)\not\prec C} = \mathbb{1}^{A_O} \otimes \tilde{W}^{A_I B_I B_O C_I C_O} + \mathbb{1}^{B_O} \otimes \tilde{W}^{A_I A_O B_I C_I C_O}, \tag{3.41}$$

where $\tilde{W}^{A_I B_I B_O C_I C_O} \geq 0$ and $\tilde{W}^{A_I A_O B_I C_I C_O} \geq 0$ are some positive semidefinite operators, whose sum gives a properly normalized process matrix. Analogously, by permutation, we obtain the form of for $W_{ecs;(A,C)\not\prec B}$ and $W_{ecs;(B,C)\not\prec A}$. Recall that an extensibly causally separable process matrix is one that is causally separable and remains so under extension of the input of the parties with entangled ancillas. We denote the closed convex cone defined by Eqs. (3.40) and (3.41), as $\mathcal{W}_{ecsep}$.

This is a wider class of process matrices, compared to the causally separable ones. Hence, one would expect that a Witness for a causally non separable process matrix, might not be able to witness the fact that a process matrix lies outside the set of extensibly causally separable process matrices. However, when witnessing the causal nonseparability of the quantum switch one would expect that restricting to the set of process matrices of the form $q W^{A \prec B \prec C} + (1 - q) W^{B \prec A \prec C}$ is enough as the general definition of extensibly causally nonseparable process matrices should reduce to this form when we restrict Charlie to be last.

Since we have the characterization of the more general tripartite extensibly causally separable process matrices, we wrote an SDP for this case, where we can witness the extensibly causal nonseparability of any tripartite process matrix. This means that we replace $\mathcal{W}_{csep}^{3C}$ by $\mathcal{W}_{ecsep}$. Out of curiosity, we searched for a witness for the quantum switch with this SDP and, curiously, we obtain different results: a different Witness with different general robustness

$$R'_r(W_{switch}) = r'^*_{switch} \approx -1.55541 \tag{3.42}$$

with visibility $v \approx 0.3913$. Recall the respective values of the previous SDP: $R_r(W_{switch}) = r^*_{switch} \approx -1.57603$, with $v \approx 0.3882$. Intuition says that these values should be the same for both SDPs; whether we allow for the closed convex set, (with which we check whether the quantum switch lies inside or outside) to have the possibility of Charlie being not last, should not affect the result because in the quantum switch Charlie is always last. This means that in the geometrical picture, we expect that along the line defined by $1°$ and W_{switch}, we expect that the two sets have the same boundary, therefore the optimal switch should have been the same. Our results suggest that the boundary is slightly different, although we do not know why.

Experimental Section

3.9 Witnessing (Experimentally) the Quantum Switch

3.9.1 Back Story

This second part of the chapter is based on Ref. [17]. We wanted to experimentally realize the quantum switch since the start of my Ph.D. To build such an experiment, we have to use two degrees of freedom of the same system (photons in our case). We spent a couple of months thinking about the various degrees of freedom that we could use, given the resources of our lab, and eventually decided on using the polarization degree of freedom of a photon as the control bit. The information carried within this degree of freedom is tranfered into path degree of freedom with a polarized beam-splitter and hence the photon takes different paths which lead to different orders of the parties A and B depending on the state of the polarization of the photon. As a target bit we decided on the Optical Angular Momentum (OAM) degree of freedom of the photon. We also decided that Alice and Bob should perform unitaries. With these constraints we sketched the setup and started to implement it in the lab. We spent some time in the lab testing our available resources to be used in the experiment, but by the time we had decided on the final sketch of the setup, we understood that the implementation of the unitary gates for Alice and Bob (inside an interferometer that is the quantum switch, as we shall see later) was a very difficult task that would be best done by our future hired experimentalist of the group. Here we report our experimental design, the measurements needed—the Witness that we obtained with our SDP programs tailored to our experimental capabilities—and our final results.

3.9.2 Witness: Taking Experimental Restrictions into Account

The first restriction that we need to implement in our SDP is the fact that the witness is composed of unitaries for A and B. A general causal witness, as we have discussed, is written in terms of the operations of the parties as follows

$$S = \sum_{ijk} \alpha_{ijk} M_i^{A_I A_O} \otimes M_j^{B_I B_O} \otimes M_k^{C_I C_O}, \tag{3.43}$$

with $d_{C_O} = 1$ in the case of the quantum switch. Therefore, in our SDP, we have to restrict our search of causal witnesses not only to in the set of valid ones, S^*, but also among those that decompose as in Eq. (3.43) where $M_i^{A_I A_O}$ and $M_j^{B_I B_O}$ are the CJ matrices of unitary operators, which we denote as $U_i^{A_I A_O}$ and $U_j^{B_I B_O}$ respectively. Hence, we are looking for witnesses of the form

$$S = \sum_{ijk} \alpha_{ijk} U_i^{A_I A_O} \otimes U_j^{B_I B_O} \otimes M_k^{C_I C_O}, \tag{3.44}$$

Now we need to find what exactly the above form means in terms of constraints on the witness S, such that we impose it on the SDP. The answer is simple: unitarity imposes constraints on the matrices $U_i^{A_I A_O}$ and $U_j^{B_I B_O}$, which imposes constraints on the operator S. These constraints are symmetry constraints on the elements of the CJ matrix (4×4 matrix) of a unitary (2×2) matrix. For a CJ matrix U of a unitary, these are the following

$$\begin{aligned} U(1, 1) &= U(4, 4) \\ U(2, 2) &= U(3, 3) \\ U(1, 2) &= -U(3, 4) \\ U(1, 3) &= -U(2, 4) \end{aligned} \tag{3.45}$$

These constraints apply to the matrices $U_i^{A_I A_O}$ and $U_j^{B_I B_O}$ and propagate to the operator S. We end up with a set of constraints for S which (together with those for a valid causal witness) define the set of witnesses that can be decomposed such that Alice and Bob perform unitaries, which we denote as $S_U \in S^*$.

We add those constraints in our SDP expressed in (3.36) where now S^* is replaced by S_U. We find the random robustness and visibility to be

$$R_{r,U} = r_U^* = -0.505836, \quad v_U \approx 0.6641 \tag{3.46}$$

Observe that having restricted our search of witnesses, we find a witness with less tolerance to noise, when compared to the optimal one with $r_{switch}^* \approx -1.57603$ and $v_{switch}^* \approx 0.3882$.

3.9.3 Implementation of the Quantum Switch

In this section we report the progress of our own implementation of the quantum switch (hence it is currently unpublished). The reason why we decided to make our own experimental realization is because one can make certain arguments that suggest a 'loophole' in the previous implementation in Refs. [11, 12]. The issue is

**Fig. 3.4 The quantum
switch, in theory**: a sketch
of one possible
implementation of the
quantum switch, where the
control qubit is encoded in
the polarization degree of
freedom. We see that
depending on the state of the
incoming photon, the two
different orders of Alice and
Bob will be realized

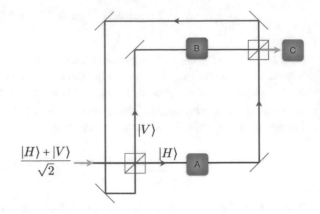

the realization of the quantum switch itself. A general, theoretical scheme of such a
realization is shown in Fig. 3.4 where the order of A and B is conditioned on the state
of the incoming photon, for example the polarization being $|H\rangle$ or $|V\rangle$. We sketch the
experimental realization of Refs. [11, 12] in Fig. 3.5. In that scheme, the issue is that
the gate that is implemented by Alice can be argued to be slightly different depending
on whether she operates first or second, although the coherence of the control bit
remains intact up to experimental error. The quantum instrument of Alice (or Bob)
that performs the unitary operation is a series of waveplates (a quarter-waveplate,
half-waveplate, quarter-waveplate configuration will perform an arbitrary unitary in
the polarization degree of freedom). If Alice is first, then the photon will go through
the series of waveplates, hitting them at some positions $\alpha_{1,2,3}$ (shown in Fig. 3.5).
Now if Alice is second, the photon will go through the same series of waveplates but
hitting them at some different position $\beta_{1,2,3}$. Hence, if one splits all waveplates in
half such that positions $\alpha_{1,2,3}$ and $\beta_{1,2,3}$ are on different halves of the waveplates, one
can see that party A is no longer one party, but two, say A_1 and A_2. In this case, if the
control bit is in state $|0\rangle$ (horizontal polarization) the order of the parties is $A_1 \prec B_1$
(and parties A_2, B_2 do not perform any operations) whereas if the control bit is $|1\rangle$
(vertical polarization) the order of the parties is $B_2 \prec A_2$. When the control bit is in
some superposition state of $|0\rangle$ and $|1\rangle$, then the two circuits are implemented 'in
superposition': $A_1 \prec B_1$ and $B_2 \prec A_2$. Clearly, this is not the implementation of the
tripartite protocol of the quantum switch, as there are five parties involved A_1, A_2,
B_1, B_2 and C.

We wanted to implement a quantum switch without this issue. In Fig. 3.6 we show
a sketch of the experimental setup. As already mentioned, the degrees of freedom
we chose are: polarization for the control bit and OAM for the target bit. In order
for Alice and Bob to implement their unitaries, each of them needs the following
elements in a sequence: two spherical lenses, two cylindrical lenses, two Dove prisms,
two cylindrical lenses, two spherical lenses. The experiment is performed with weak
coherent pulses, prepared in a diagonal polarization state $|D\rangle$ and in the 00 OAM
mode (TEM$_{00}$). Charlie, sitting at the end of the experiment, makes a measurement

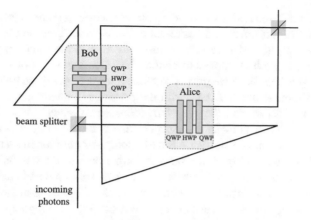

Fig. 3.5 The quantum switch, as implemented in Ref. [11]: a sketch of the experimental setup where the control qubit is encoded in the path degree of freedom of the incoming photon. Depending on the output port of the initial beam splitter, the photon will go through Alice and Bob with different orders

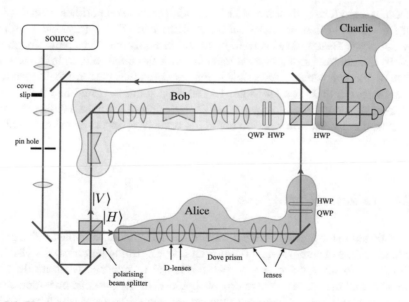

Fig. 3.6 The quantum switch, as implemented in our experiment: a sketch of the experimental setup where the control qubit is encoded in the polarization degree of freedom of the incoming photon. Depending on the output port of the initial polarized beam splitter, the photon will go through Alice and Bob with different orders

on the control bit in the D/A basis, according to the Witness, which we discuss below. In this implementation, it is no longer the case that the parties can be 'split' in two, as the incoming system to the parties A and B is coming from the same spatial position. However, one can make a similar argument about each party that may perform a different unitary depending on the incoming polarization. It is known that Dove prisms perform slightly different unitaries depending on the polarization. Hence, from a position degree of freedom (which was the issue on the previous implementations) we have moved to another issue, less obvious or experimentally significant, regarding the Dove prisms' performance depending on polarization.

Moreover, there is another issue with the previous implementation that is fixed in our case: just as the path degree of freedom was problematic in the previous implementations, one can make the same argument about the time of arrival of the photons at each station. The fact that A is performing its operation in a 'superposition' of now and later (depending on the short or longer path of the photon) could make someone ask 'do the waveplates behave in the same way now and later?' A way to tackle such an argument is to use long photons—at least longer than the length of the experimental setup. We used weak coherent pulses to detect single photons, with a coherence length much larger that our setup. This results in Alice and Bob operating on their incoming photon over a long period of time (because the photon is largely delocalized) such that the difference between 'now and later' will be not be an issue anymore.

Despite our efforts to fix some of the issues with previous experimental realizations of the quantum switch, we have not reached the end of the story. Even with the previous experimental issues solved, there still remains one conceptual 'loophole'. Namely, the control bit that travels together with the target bit: so long there is a degree of freedom which is accessible to the parties as they perform their operations on the target bit, there is a loophole. Therefore, a solution would be for them (the degrees of freedom) to be separate, namely for the control bit to be a different system than the target bit. However, we do not know if such an experimental realization is possible.

3.9.4 Tailoring Our Witness

As mentioned in the previous section, we obtained a Witness, suitable for this implementation of the quantum switch (with A and B performing unitaries). Its visibility, with respect to white noise was $v_U \approx 0.6641$. Let us now see what are the exact operations that the parties have to do. We begin with the possible operations, such that we have a set of combinations of the parties operations from which we can find the different combinations and weights in Eq. (3.43). Charlie can measure 4 possible observables of the control bit: $\{\mathbb{1}, X, Y, Z\}$ with X, Y, Z being the Pauli matrices: $\sigma_x, \sigma_y, \sigma_z$. The CJ matrices of the unitaries of Alice and Bob live in a $4 \times 4 - 6 = 10$-dimensional space. Note that the constraints in Eq. (3.45) are 6, because the last two yield two for the real part and two for the imaginary. This gives the dimension of S_U: initially S is a 32×32 matrix ($d_{A_I} d_{A_O} d_{B_I} d_{B_O} d_{C_I} = 32$), which says that it lives in a 1024-dimensional space, and given the constraints upon S_U, which are 624 in total, we have that S_U lives in a 400-dimensional space.

The fact that $U_i^{A_I A_O}$ and $U_j^{B_I B_O}$ are 10-dimensional means that there are 10 unitaries that span the Hilbert space of the operations of A and B. Hence, there are 10 possible unitaries for these parties. Together with the fact that Charlie has 4 different possible operations, we may end up with 400 different terms in S_U, which means that we may (depending on the amount of no-zero coefficients) have to run our experiment 400 times to obtain all the terms in S_U. For this reason, we decided to restrict the amount of unitaries. We chose the following unitaries, given that any unitary can be written as a combination of Pauli matrices, $\{I, X, Y, Z, (X - Z)/\sqrt{2}, (X + Z)/\sqrt{2}, (Y + Z)/\sqrt{2}\}$. Restricting S_U to be composed of unitaries for Alice and Bob only from this set, we obtain a Witness with $r^* = -0.248225$ and $v = 0.8011$. We have significantly decreased the tolerance of our witness to white noise, but the number of possible measurements is now 144 and after calculating the exact coefficients we find that they are always zero when Charlie performs Y, Z and sometimes zero on all other cases. The exact number of terms in the Witness is now 47. This means that 47 measurements need to be performed in the lab, each measurement consisting of a unitary operation for Alice, another one for Bob, and a measurement of the control bit for Charlie. The latter measurement outcome will provide the statistics needed.

3.10 Conclusion

In this chapter, we brought the process matrix framework and all its oddities one step closer to the reality of an experimentalist. The framework is predicting scenarios incompatible with a definite causal order between some parties, and some of those scenarios have the capability to produce noncausal correlations; i.e violating causal inequalities.

In the theoretical part of this chapter, based on Ref. [1], we provided a way to prove the causal nonseparability of any given process matrix. This is done through a theoretical object, called causal witness. Keeping an eye on experimental realizations of causally nonseparable process matrices, this causal witness (as with entanglement witnesses) corresponds to measurements that need to be done in the lab. To obtain a witness, one needs to characterize the set of causally separable process matrices. We provide such a characterization for the general bipartite case and a special tripartite case. We provide the SDP that looks for the optimal causal witness for a given process matrix in these cases. This optimality is in terms of tolerance to white noise (represented by a joint mixed state that all the parties receive) although it can be altered to represent tolerance to any kind of noise model, depending on the experimental realization. We applied our results to the quantum switch: a tripartite causally nonseparable process matrix, which has an experimental realization. We obtained its optimal causal witness.

In the experimental part of this chapter, based on Ref. [17], we tailor even further our theoretical results, to meet the requirements of our own experimental realization of the quantum switch. We obtain a causal witness that allows for the two parties to perform only unitaries, and even restrict those ones to only seven. The third party performs a projective measurement. Finally, we briefly present the experimental setup, where the target qubit is encoded in the OAM degree of freedom and the control qubit in the polarization degree of freedom. We comment on the advantages over similar past experiments and justify our excitement for our implementation of the quantum switch.

Appendix: Process Matrix Characterization

Bipartite valid process matrices: Here we provide the derivation of the Eq. (3.5) that completely characterize the set of bipartite causally separable process matrices.

The set of valid process matrices is defined by the following requirements: the bipartite expression of the probabilities in Eq. (3.1) must be non-negative and sum up to one for all possible operations (including operations that act on ancillary systems on arbitrary input states, which we ignore at the moment although it leads to the same conclusion). These yield

$$W \geq 0 \tag{3.47}$$

$$\text{Tr}[(M^{A_I A_O} \otimes M^{B_I B_O})W^{A_I A_O B_I B_O}] = 1 \quad \forall M^{A_I A_O} \geq 0, \ M^{B_I B_O} \geq 0, \tag{3.48}$$

$$\text{s.t.} \quad {}_{A_O} M^{A_I A_O} = \mathbb{1}^{A_I A_O}/d_{A_O}, \ {}_{B_O} M^{B_I B_O} = \mathbb{1}^{B_I B_O}/d_{B_O}. \tag{3.49}$$

We remind the definition of two maps, we will use extensively

$$_x W = \frac{\mathbb{1}^x}{d_x} \text{Tr}_x W \tag{3.50}$$

$$_{[1-x]} W = W - {}_x W, \tag{3.51}$$

Using the second map we defined above, we can see that for any two Hermitian operators X and Y, the operators ${}_{[1-A_O]}X + \mathbb{1}^{A_I A_O}$ and ${}_{[1-B_O]}Y + \mathbb{1}^{B_I B_O}$ satisfy the above normalization constraints in Eq. (3.49). Then we write the normalization condition of the probabilities of Eq. (3.48) as

$$\text{Tr}[({}_{[1-A_O]}X + \mathbb{1}^{A_I A_O} \otimes {}_{[1-B_O]}Y + \mathbb{1}^{B_I B_O} W] = 1, \quad \forall X, Y \tag{3.52}$$

For $X = Y = 0$ this yields

$$\text{Tr}(W) = d_{A_O} d_{B_O}. \tag{3.53}$$

For $Y = 0$ and $X = 0$ in turn, implies

$$\text{Tr}[({}_{[1-A_O]}X \otimes \mathbb{1}^{B_I B_O} W] = 0, \quad \forall X,$$
$$\text{Tr}[(\mathbb{1}^{A_I A_O} \otimes {}_{[1-B_O]} Y)W] = 0, \quad \forall Y, \tag{3.54}$$

because for $Y = 0$ we have

$$\text{Tr}[(_{[1-A_O]}X + \mathbb{1}^{A_I A_O}) \otimes \mathbb{1}^{B_I B_O})W] = 1 \Rightarrow$$

$$\text{Tr}[(_{[1-A_O]}X \otimes \mathbb{1}^{B_I B_O})W] + \text{Tr}[(\mathbb{1}^{A_I A_O}) \otimes \mathbb{1}^{B_I B_O})W] = 1 \Rightarrow$$

$$\text{Tr}[(_{[1-A_O]}X \otimes \mathbb{1}^{B_I B_O})W] = 0$$

Back to the Eq. (3.54), they imply

$$\text{Tr}[(_{[1-A_O]}X \otimes_{[1-B_O]} Y)W] = 0 \tag{3.55}$$

Finally, thinking of the trace as a Hilbert-Schmidt inner product and the fact that the maps $_{[1-A_O]}$ and $_{[1-B_O]}$ are self-adjoint, the above conditions (3.54), (3.55) are equivalent to

$$_{[1-A_O]}(\text{Tr}_{B_I B_O} W) = 0$$
$$_{[1-B_O]}(\text{Tr}_{A_I A_O} W) = 0 \tag{3.56}$$
$$_{[1-A_O][1-B_O]}W = 0.$$

which we rewrite as

$$_{B_I B_O}W = _{A_O B_I B_O}W$$
$$_{A_I A_O}W = _{A_I A_O B_O}W \tag{3.57}$$
$$W = _{B_O}W + _{A_O}W - _{A_O B_O}W$$

Each of these conditions defines a linear subspace, whose intersection is a subspace on which all valid bipartite process matrices live, and we denote as

$$\mathcal{L}_V = \{W \in A_I \otimes A_O \otimes B_I \otimes B_O | W = L_V(W)\}, \tag{3.58}$$

The projector onto this subspace, L_V will be used in many occasions in this chapter, hence it is useful to find its expression. Firstly, we can write each of the equations in (3.57) as a projector onto a subspace

$$W = L_A(W), \quad W = L_B(W), \quad W = L_{AB}(W) \tag{3.59}$$

where the projectors are

$$L_A = W - _{B_I B_O}W + _{A_O B_I B_O}W,$$
$$L_B = W - _{A_I A_O}W + _{A_I A_O B_O}W, \tag{3.60}$$
$$L_{AB} = _{B_O}W + _{A_O}W - _{A_O B_O}W$$

Then the projector we are looking for, L_V, is the intersection of the subspaces of the above three projectors, and is given simply by their composition, i.e. $L_V(W) = L_A \circ L_B \circ L_{AB}$, which can be written as

$$L_V(W) = {}_{A_O}W + {}_{B_O}W - {}_{A_O B_O}W - {}_{B_I B_O}W + {}_{A_O B_I B_O}W - {}_{A_I A_O}W + {}_{A_I A_O B_O}W \tag{3.61}$$

This completes the characterization of the set of bipartite process matrices: an operator $W \in A_I \otimes A_O \otimes B_I \otimes B_O$ is valid if and only if $W \geq 0$, $\mathrm{Tr}\, W = d_{A_O} d_{B_O}$ and $W = L_V(W)$.

Tripartite valid process matrices: Here we provide the characterization of the set of valid process matrices in the tripartite case and write the explicit expression of the projector onto the subspace they live. An analogous to the bipartite case argument leads to the following conclusion: an operator $W \in A_I \otimes A_O \otimes B_I \otimes B_O \otimes C_I \otimes C_O$ is a valid tripartite process matrix if and only if $W \geq 0$, $\mathrm{Tr}\, W = d_{A_O} d_{B_O} d_{C_O}$ and

$$
\begin{aligned}
W &= L_X(W), \quad X = \{A, B, C\} \\
W &= L_{XY}(W), \quad \{X, Y\} = \{A, B, C\} \\
W &= L_{ABC}(W)
\end{aligned}
\tag{3.62}
$$

where the maps $L_A, L_B, L_C, L_{AB}, L_{AC}, L_{BC}, L_{ABC}$ are commuting projectors onto linear subspaces of $A_I \otimes A_O \otimes B_I \otimes B_O \otimes C_I \otimes C_O$ defined by

$$
\begin{aligned}
L_A(W) &= {}_{[1-(1-A_O)B_I B_O C_I C_O]}W\,, \\
L_B(W) &= {}_{[1-(1-B_O)A_I A_O C_I C_O]}W\,, \\
L_C(W) &= {}_{[1-(1-C_O)A_I A_O B_I B_O]}W\,, \\
L_{AB}(W) &= {}_{[1-(1-A_O)(1-B_O)C_I C_O]}W\,, \\
L_{AC}(W) &= {}_{[1-(1-A_O)(1-C_O)B_I B_O]}W\,, \\
L_{BC}(W) &= {}_{[1-(1-B_O)(1-C_O)A_I A_O]}W\,, \\
L_{ABC}(W) &= {}_{[1-(1-A_O)(1-B_O)(1-C_O)]}W
\end{aligned}
\tag{3.63}
$$

where we used the shorthand notation

$$
{}_{[\sum_X \alpha_X X]}W = \sum_X \alpha_X \cdot {}_X W
\tag{3.64}
$$

for a sum over products of subsystems X with coefficients α_X (and with ${}_1 W := W$).

The above constraints are equivalent to $W = L_V(W)$, where the map L_V is obtained by composing the 7 maps in (3.62), and is expressed as

$$
L_V(W) = {}_{\left[1-(1-A_O+A_I A_O)(1-B_O+B_I B_O)(1-C_O+C_I C_O)+\, A_I A_O B_I B_O C_I C_O\right]}W
\tag{3.65}
$$

which defines a projector onto the linear subspace

$$\mathcal{L}_V = \left\{ W \in A_I \otimes A_O \otimes B_I \otimes B_O \otimes C_I \otimes C_O \mid W = L_V(W) \right\}. \qquad (3.66)$$

This completes the characterization of the valid tripartite process matrices. For the n-partite case, refer to the Appendix B of the relevant paper [1].

References

1. Araújo M et al (2015) Witnessing causal nonseparability. New J Phys 17:102001
2. Oreshkov O, Costa F, Brukner Č (2012) Quantum correlations with no causal order. Nat Commun 3:1092
3. Oreshkov O, Giarmatzi C (2016) Causal and causally separable processes. New J Phys 18:093020
4. Hardy L (2009) Foliable operational structures for general probabilistic theories. In: Halvorson H (ed) Deep beauty: understanding the quantum world through mathematical innovation, p 409
5. Coecke B (2010) Quantum picturalism. Contemp Phys 51:59–83
6. Chiribella G, D'Ariano GM, Perinotti P (2010) Probabilistic theories with purification. Phys Rev A 81:062348
7. Chiribella G, D'Ariano GM, Perinotti P (2011) Informational derivation of quantum theory. Phys Rev A 84:012311
8. Chiribella G, D'Ariano GM, Perinotti P, Valiron B (2013) Quantum computations without definite causal structure. Phys Rev A 88:022318
9. Chiribella G (2012) Perfect discrimination of no-signalling channels via quantum superposition of causal structures. Phys Rev A 86:040301
10. Araújo M, Costa F, Brukner Č (2014) Computational advantage from quantum-controlled ordering of gates. Phys Rev Lett 113:250402
11. Procopio LM et al (2015) Experimental superposition of orders of quantum gates. Nat Commun 6.7913
12. Rubino G et al (2017) Experimental verification of an indefinite causal order. Sci Adv 3
13. Rockafellar RT (1970) Convex analysis. Princeton University Press
14. Gutoski G, Watrous J (2006) Toward a general theory of quantum games. In: Proceedings of 39th ACM STOC, pp 565–574
15. Chiribella G, D'Ariano GM, Perinotti P (2009) Theoretical framework for quantum networks. Phys Rev A 80:022339
16. Nesterov Y, Nemirovskii A (1987) Interior point polynomial algorithms in convex programming. Studies in Applied Mathematics (Society for Industrial and Applied Mathematics)
17. Goswami K et al (2018) Indefinite causal order in a quantum switch. Phys Rev Lett 121:090503

Chapter 4
Causal Polytopes

4.1 Back Story

In Chap. 2, causal and causally separable processes, we saw two different descriptions of the same situation. The situation is that there are a number of parties operating on their incoming system, obtaining an outcome, and send out an outgoing system. As we have seen we describe this situation through the process matrix formalism, assuming that the systems and the operations on them are described by quantum mechanics—this description is device-dependent. Another description, device-independent, is at the level of probabilities of the outcomes of the parties, given their settings. This is in complete analogy with quantum correlations: we can describe the situation using a state (like a bipartite entangled state) or using the obtained correlations, after performing measurements on the state. Then we can classify the states into separable and nonseparable states and the correlations into local or nonlocal correlations—depending on whether they satisfy a set of constraints. Local correlations form a convex polytope.

Similarly, in Chap. 2 we discussed that the probabilities that respect causality (as defined in the Chapter) form a convex polytope. However, a rigorous proof was lacking. Furthermore, characterization of a particular causal polytope, defined by a number of parties and with given settings and outcomes, is an interesting task in its own right. This is because the facets of the polytope are causal inequalities, that is, restrictions on the correlations that satisfy causality. As we know, the process matrix formalism allows for violations of such causal inequalities, hence this line of investigation can lead to new ways (process matrices) to observe violations of causal inequalities, even though an experimental observation of such extraordinary results is still lacking.

After discussions with two collaborators, experts in convex polytope characterization, we decided on the project: to lay down the tools for characterization of causal polytopes and investigate some particular cases. Using our previous knowledge of the concepts around the process matrix formalism [1] and of those developed in Chap. 2 and Ref. [2], we defined causal correlations, provided some of its properties,

© Springer Nature Switzerland AG 2019
C. Giarmatzi, *Rethinking Causality in Quantum Mechanics*, Springer Theses,
https://doi.org/10.1007/978-3-030-31930-4_4

proved rigorously that they form a convex polytope, and finally studied a particular tripartite causal polytope. In this Chapter, we describe the ideas developed for this project and the main results, both reported in Ref. [3]. We also describe the specific technique for a bipartite polytope characterization.

4.2   Introduction

The study of correlations between a number of measurement stations is of course not new. Bell correlations are the most famous ones, that arise from two distant measurement stations that receive an entangled state [4]. The counter-intuitive results of such measurements required a systematic way of studying quantum correlations, i.e. to classify them in terms of certain constraints that they should satisfy (a Bell-type inequality). In our case, we are interested in classifying correlations between a number of parties, that satisfy a notion of causality—which translates to similar inequalities. As in the case of Bell-local correlations, for a fixed number of settings and outcomes for the parties, the *causal* correlations form a *causal* polytope whose non-trivial facets define *causal inequalities*.

Studying correlations is closely connected to studying causal structures: for example, the no-signaling constraints (Bell-type inequalities) that local correlations should obey, correspond to constraints in the causal structure of the experiments (their correlations are due to a common cause). There is a growing interest in studying and characterizing general causal structures [5–8] and understanding when the corresponding correlations can be produced with classical or quantum systems [9, 10]. Here we are interested in correlations that can be achieved by a well-defined causal structure, i.e. one that obeys causality, and the corresponding constraints on the correlations are the causal inequalities

The interest in causal inequalities comes from the process matrix formalism [1], which, very surprisingly, allows for scenarios that violate such causal inequalities. To remind you of the core of that formalism: it is assumed that a number of parties are localized inside closed laboratories that receive a (quantum) system, the party performs a quantum operation on that system obtaining an outcome, and then (possibly conditionally on the outcome) sends another system out. The process matrix formalism provides a characterization of the allowed correlations (joint probabilities of the local outcomes given their settings) *without* any assumption on the global causal structure in which the parties are embedded. Assuming a notion of causality (like the one in Chap. 2) one can see how it manifests in terms of the description of a given scenario: at the level of probabilities (causal correlations), or at the level of the process matrix.

The interesting part is that process matrices have been found that can be used by the parties as a resource to give rise to noncausal correlations [1]. Unfortunately there has been no experimental realization of such scenarios—even such a possibility is not certain. However, we can still develop the tools for the study of causal polytopes and causal inequalities. In this chapter, based on Ref. [3], we define causal correlations:

joint probabilities of the local outcomes of the parties, which we will call *inputs*, given their settings, which we call *outputs*. Our definition follows the ones of Ref. [2], on which Chap. 2 is based. We show some properties of these correlations. We prove that for a fixed number of parties and their inputs and outputs, the correlations form a convex polytope whose vertices correspond to *deterministic strategies*. This means that any causal correlation can be expressed as a probabilistic mixture of deterministic causal strategies. Deterministic causal strategies are those where the output of each party is a deterministic function of its input and of the input of the parties that have operated in its past. We describe the specifics of polytope characterization for a given scenario: from obtaining the vertices, to providing them as an input to some software to obtain the facets of the polytope. For that we use a simple bipartite case. We apply this technique to analyze the simplest tripartite non-trivial case in terms of its causal polytope and its nontrivial facets. We provide the main results of this analysis and finally we discuss our results.

4.3 Multipartite Causal Correlations

4.3.1 Some Notations

In our study, we consider situations where a finite number of parties is involved, $N \geq 1$, each named A_k, $k = 1, \cdots, N$. Each party A_k obtains an input x_k from some finite set and generate an output a_k also from some finite set (different in principle for each x_k). (Note that although in different Chapters we call x_k a setting and a_k an outcome, here we use the convention of calling them inputs and outputs as this is common practice in similar analysis of correlations in terms of polytope characterization.) We call a 'scenario': the fixed number of parties, the set of possible inputs for each party, and the set of possible outputs for each input for each party. We define the vectors of inputs and outputs $\vec{x} = (x_1, \ldots, x_N)$ and $\vec{a} = (a_1, \ldots, a_N)$. The probabilities (or correlation) obtained by N parties in a fixed scenario is then described by the conditional probability distribution $P(\vec{a}|\vec{x})$.

We denote the full set of parties as, $\mathcal{N} = \{1, 2, \cdots, N\}$, a subset of parties $\mathcal{K} \subset \mathcal{N}$, as $\mathcal{K} = \{k_1, k_2, \cdots, k_K\}$. The lists of inputs and outputs for these K parties are $\vec{x}_{\mathcal{K}} = (x_{k_1}, x_{k_2}, \cdots, x_{k_K})$ and $\vec{a}_{\mathcal{K}} = (a_{k_1}, a_{k_2}, \cdots, a_{k_K})$. This allows us to write marginal correlations: for example, for the subset $\mathcal{K}$, we have $P(\vec{a}_{\mathcal{K}}|\vec{x}) = \sum_{\vec{a}_{\mathcal{N}\setminus\mathcal{K}}} P(\vec{a}|\vec{x})$. When $\mathcal{K} = \{k\}$, where we single out one party, A_k, we simply write k instead of $\mathcal{K}$ and we write the vectors for the remaining $N - 1$ parties in $\mathcal{N}\setminus k$ as $\vec{x}_{\setminus k} = (x_1, \cdots, x_{k-1}, x_{k+1}, \cdots, x_N)$ and $\vec{a}_{\setminus k} = (a_1, \cdots, a_{k-1}, a_{k+1}, \cdots, a_N)$.

4.3.2 Definition and Properties

The definition of causal correlations in the multipartite case (referred to as causal processes) is one of the main results in Chap. 2 and Ref. [2]. Here we adapt this definition (in particular, the one referred to as 'canonical causal decomposition') to describe causal correlations for a number of parties that are embedded in a well-defined causal structure, be it probabilistic or dynamical. In this definition, we are not concerned with the causal order of the parties, we simply write the correlations in terms of probability of inputs given outputs. We also denote dependence of some correlation $P(\vec{a}|\vec{x})$ on a variable b, as $P_b(\vec{a}|\vec{x})$.

Definition 5 (*Multipartite causal correlations*)

- For $N = 1$, any valid probability distribution $P(a_1|x_1)$ is *causal*;
- For $N \geq 2$, an N-partite correlation is *causal* if and only if it can be decomposed in the form

$$P(\vec{a}|\vec{x}) = \sum_{k \in \mathcal{N}} q_k \, P_k(a_k|x_k) \, P_{k,x_k,a_k}(\vec{a}_{\backslash k}|\vec{x}_{\backslash k}), \tag{4.1}$$

with $q_k \geq 0$ for each k, $\sum_k q_k = 1$, where (for each k) $P_k(a_k|x_k)$ is a single-party (and hence causal) probability distribution and (for each k, x_k, a_k) $P_{k,x_k,a_k}(\vec{a}_{\backslash k}|\vec{x}_{\backslash k})$ is a causal $(N-1)$-partite correlation.

We move on to some properties of such causal correlation. The proofs of these properties are in the Appendix of the related paper [3].

Convexity: Any convex mixture of causal correlations, for a given scenario, is also a causal correlation.

Ignoring (the outputs of) some parties: Any marginal correlation, for any subset of parties, of a causal correlation is also causal. In particular, consider an N partite causal correlation $P(\vec{a}|\vec{x})$ and a (nonempty) subset $\mathcal{K} \subset \mathcal{N}$ with K parties. Then the K-partite correlation

$$P_{\vec{x}_{\mathcal{N}\backslash\mathcal{K}}}(\vec{a}_{\mathcal{K}}|\vec{x}_{\mathcal{K}}) := P(\vec{a}_{\mathcal{K}}|\vec{x}) = \sum_{\vec{a}_{\mathcal{N}\backslash\mathcal{K}}} P(\vec{a}|\vec{x}) \tag{4.2}$$

is causal for all $\vec{x}_{\mathcal{N}\backslash\mathcal{K}}$.

Note that the correlation written above, $P_{\vec{x}_{\mathcal{N}\backslash\mathcal{K}}}(\vec{a}_{\mathcal{K}}|\vec{x}_{\mathcal{K}})$, is conditioned only on the inputs of the $N - K$ parties and not the outputs, as denoted by the subscript $\vec{x}_{\mathcal{N}\backslash\mathcal{K}}$. This is the case when the outputs of the parties are discarded. Hence, when a set of parties have a causal correlation, when the outputs are discarded for some subset of the parties, the marginal correlation for the remaining parties would also be causal. If the outputs of the $N - K$ parties are not discarded, then the above correlation would be $P_{\vec{x}_{\mathcal{N}\backslash\mathcal{K}},\vec{a}_{\mathcal{N}\backslash\mathcal{K}}}(\vec{a}_{\mathcal{K}}|\vec{x}_{\mathcal{K}})$ and there is no guarantee that their correlation would be causal.

Note that if the outputs of the $N - K$ parties are not discarded, then the correlation for the K parties is conditioned on the inputs and outputs of the $N - K$ parties. A correlation conditional on some outputs of some parties implies post-selection, which is known to be able to turn causal into noncausal correlations [2, 11].

Combining causal correlations 'one after the other': If for two sets of parties $\mathcal{K}$, $\mathcal{N}\backslash\mathcal{K}$, it is that $\mathcal{K} \prec \mathcal{N}\backslash\mathcal{K}$, in the sense that all the parties in $\mathcal{K}$ are acting before the parties in $\mathcal{N}\backslash\mathcal{K}$, and if each sets are described by a causal correlation, then the correlation for the combined set, i.e.

$$P(\vec{a}|\vec{x}) := P(\vec{a}_{\mathcal{K}}|\vec{x}_{\mathcal{K}}) \, P_{\vec{x}_{\mathcal{K}},\vec{a}_{\mathcal{K}}}(\vec{a}_{\mathcal{N}\backslash\mathcal{K}}|\vec{x}_{\mathcal{N}\backslash\mathcal{K}}), \tag{4.3}$$

is also causal.

An equivalent form of multipartite causal correlations: For the case of $N \geq 2$, we find an equivalent form for an N-partite causal correlation

$$P(\vec{a}|\vec{x}) = \sum_{\emptyset \subsetneq \mathcal{K} \subsetneq \mathcal{N}} q_{\mathcal{K}} \, P_{\mathcal{K}}(\vec{a}_{\mathcal{K}}|\vec{x}_{\mathcal{K}}) \, P_{\mathcal{K},\vec{x}_{\mathcal{K}},\vec{a}_{\mathcal{K}}}(\vec{a}_{\mathcal{N}\backslash\mathcal{K}}|\vec{x}_{\mathcal{N}\backslash\mathcal{K}}) \tag{4.4}$$

(note the P is a different function for each subset $\mathcal{K}$) where the sum runs over all nonempty strict subsets of $\mathcal{N} = \{1, 2, \cdots, N\}$, with $q_{\mathcal{K}} \geq 0$ for each subset $\mathcal{K}$, $\sum_{\mathcal{K}} q_{\mathcal{K}} = 1$. Also, $\mathcal{K}$, $P_{\mathcal{K}}(\vec{a}_{\mathcal{K}}|\vec{x}_{\mathcal{K}})$ is a $|\mathcal{K}|$-partite correlation and (for each $\mathcal{K}, \vec{x}_{\mathcal{K}}, \vec{a}_{\mathcal{K}}$), $P_{\mathcal{K}}(\vec{a}_{\mathcal{K}}|\vec{x}_{\mathcal{K}}) \, P_{\mathcal{K},\vec{x}_{\mathcal{K}},\vec{a}_{\mathcal{K}}}(\vec{a}_{\mathcal{N}\backslash\mathcal{K}}|\vec{x}_{\mathcal{N}\backslash\mathcal{K}})$ is an $N - |\mathcal{K}|$-partite correlation.

This characterization is reminiscent to the original definition in Chap. 2 and Ref. [2], with the difference that there, the $\mathcal{K}$-partite causal correlation was also a nonsignaling correlation.

This equivalent characterization implies a correlation of this form is causal (but as we will see later, not every causal correlation can be written in this form), where the subset $\mathcal{K}$ contains all of $\mathcal{N}$ except one party, labelled k. Hence, the following correlation is causal:

$$P(\vec{a}|\vec{x}) = \sum_{k \in \mathcal{N}} q_k \, P_k(\vec{a}_{\backslash k}|\vec{x}_{\backslash k}) \, P_{k,\vec{x}_{\backslash k},\vec{a}_{\backslash k}}(a_k|x_k), \tag{4.5}$$

(remember the subscript denotes dependence, not labelling of the correlation) with $q_k \geq 0$, for each k, $\sum_k q_k = 1$, where (for each k) $P_k(\vec{a}_{\backslash k}|\vec{x}_{\backslash k})$ is an $(N - 1)$-partite causal correlation, and (for each k, $\vec{x}_{\backslash k}$, $\vec{a}_{\backslash k}$), $P_{k,\vec{x}_{\backslash k},\vec{a}_{\backslash k}}(a_k|x_k)$ is a single-party probability distribution. The intuitive picture behind this form is that, while the original definition decomposes a causal correlation into terms where one party is 'first', the latter form distinguishes a party that is 'last'. Note that the correlations arising from the *quantum switch* described in Chaps. 2 and 3 are of precisely that form and hence causal, as proven in Refs. [2, 11]. However, for the case three parties or more, $N \geq 3$, the above form is not sufficient for a correlation to be causal. We provide below some examples of such non-equivalence.

Some Examples

Example 1 Our first example of a causal correlation, is one that is compatible with a fixed causal order between the parties; i.e. there is a variable defining the causal order of the parties that is chosen in advance and independently of all parties inputs and outputs. For example, a correlation compatible with the causal order $A_1 \prec A_2 \prec \cdots A_N$ can be written as

$$P(\vec{a}|\vec{x}) = P(\vec{a}_1|\vec{x}_1)\, P_{x_1,a_1}(\vec{a}_2|\vec{x}_2)\, P_{x_1,a_1,x_2,a_2}(\vec{a}_3|\vec{x}_3) \times \cdots \times P_{\vec{x}_{\backslash N},\vec{a}_{\backslash N}}(\vec{a}_N|\vec{x}_N), \quad (4.6)$$

which of course satisfies the definition of a causal correlation in Eq. (4.1).

Example 2 The next step towards a more general example is to consider probabilistic mixtures of the above situation. The resulting correlation is of course causal, as we have discussed above on the convexity of causal correlations. Consider the following situation: with probability q the correlation P is compatible with the causal order $A_{s(1)} \prec A_{s(2)} \prec \cdots \prec A_{s(N)}$ and takes the value P_s, and with probability $(1-q)$ the correlation is compatible with the causal order $A_{t(1)} \prec A_{s(2)} \prec \cdots \prec A_{t(N)}$ and takes the value P_t (where s and t are two permutations of $\{1, 2, \cdots, N\}$). Then, for any $q \in [0, 1]$, the correlation is written

$$P = qP_s + (1-q)P_t \qquad (4.7)$$

and it is also causal.

Example 3 Taking one step further towards more general situations, we require three parties at least. This is the case of dynamical causal order Ref. [2] where the causal order of some parties may depend on the inputs and outputs of some other parties in their past. Consider the following conditions: three parties, each with binary inputs, A_1 has a single fixed output (which we can therefore ignore), and parties A_2, A_3 have binary outputs, and the particular (deterministic) correlation,

$$P(a_1, a_2|x_1, x_2, x_3) = \delta_{x_1,0}\, \delta_{a_2,0}\, \delta_{a_3,x_2} + \delta_{x_1,1}\, \delta_{a_2,x_3}\, \delta_{a_3,0}, \qquad (4.8)$$

where δ is the Kronecker delta. This correlation is tailored such that it describes the following situation: observing the factor $\delta_{x_1,i}$, $i = 0, 1$ in each of these terms we understand that the correlation for the parties A_2, A_3 depends on the input of A_1—hence we can think that A_1 acts first and depending on their choice for x_1, the correlation for the remaining parties changes. For $x_1 = 0$ the first term is 'activated', which tells us that A_2 outputs 0 and $a_3 = x_2$, which means that A_2 communicates their input to A_3, compatible with $A_2 \prec A_3$. Now if the input of A_1 is $x_1 = 0$, the second term gets activated, which is compatible with $A_3 \prec A_2$. Therefore, we see that A_1 can 'select' between the two possible causal configuration for the parties

A_2, A_3. As we discussed in Chap. 2, this situation is compatible with a well-defined, albeit dynamical, causal order between the parties for each run of the experiment. Note that this correlation, although causal, is not of the form of Eq. (4.5); indeed there is no party that always acts last.

Example 4 The final example is a generalization of the example just above. The order of A_2, A_3 depends on the input of A_1 like before, but more generally. Their order is a probabilistic function of the input of A_1; that is, different values of x_1 'activate' different probabilities for one order to be realized or the other. Let us see a concrete example

$$
\begin{aligned}
p(a_2, a_3 | x_1, x_2, x_3) = {} & \delta_{x_1,0} \left(q_0 \, \delta_{a_2,0} \, \delta_{a_3,x_2} + (1 - q_0) \, \delta_{a_2,x_3} \, \delta_{a_3,0} \right) \\
& + \delta_{x_1,1} \left(q_1 \, \delta_{a_2,0} \, \delta_{a_3,x_2} + (1 - q_1) \, \delta_{a_2,x_3} \, \delta_{a_3,0} \right)
\end{aligned}
\tag{4.9}
$$

with $\{q_0, q_1\} \in (0, 1)$ (and not $\in [0, 1]$ because that would be the deterministic case). We can see that if $x_1 = 0$ the first term describes that $A_2 \prec A_3$ with probability q_0, while $A_3 \prec A_2$ with probability $(1 - q_0)$. For $x = 1$, the second term prevails and now we have $A_2 \prec A_3$ with probability q_1 and $A_3 \prec A_2$ with probability $(1 - q_1)$. This correlation is of course causal and it is indeed of the form of Eq. (4.1).

4.4 Further Investigation of Causal Correlations

As we discussed already, any convex combination of causal correlations is also causal. This means that for a given scenario (fixed number of parties, each with finite sets of possible inputs and outputs), causal correlations form a convex set. In fact it has been argued (in Chap. 2, Refs. [2, 12], and was proven for the bipartite case in Ref. [12]) that it is a convex polytope, the so-called *causal polytope*. In this section we prove that any causal correlation can be written as a convex combination of deterministic causal correlations. The polytope structure follows from the fact that, for any given scenario, these deterministic causal correlations are finite; so any causal correlation can be described in terms of a finite set of deterministic causal correlations that represent the vertices of the polytope. The facets of the causal polytope represent linear inequalities that are satisfied by all causal correlations (the points inside the polytope). The nontrivial inequalities (we will see what this means later) correspond to ('tight' as they are called) *causal inequalities*.

4.4.1 Decomposing Causal Correlations into Deterministic Ones

Here comes some more notations: We call a correlation deterministic if the list of outputs $\vec{a}$ is a deterministic function $\vec{\alpha}$ of the list of inputs, $\vec{x} : \vec{a} = \vec{\alpha}(\vec{x})$. We denote this deterministic probability distribution as

$$P_{\vec{\alpha}}^{\text{det}}(\vec{a}|\vec{x}) = \delta_{\vec{a},\vec{\alpha}(\vec{x})} \tag{4.10}$$

To prove that causal correlations form a polytope, we prove the following theorem.

Theorem 6 *Any N-partite causal correlation can be written as a convex combination of a finite number of deterministic causal correlations*

$$P(\vec{a}|\vec{x}) = \sum_{\vec{\alpha}} q_{\vec{\alpha}} \, P_{\vec{\alpha}}^{\text{det}}(\vec{a}|\vec{x}) \tag{4.11}$$

with $q_{\vec{\alpha}} \geq 0$, $\sum_{\vec{\alpha}} q_{\vec{\alpha}} = 1$, where the sum is over all functions $\vec{\alpha} : \vec{x} \mapsto \vec{a}$ that define a deterministic causal correlation $P_{\vec{\alpha}}^{\text{det}}(\vec{a}|\vec{x})$.

The proof is by induction:

- For $N = 1$, we know that any correlation can be written as a convex combination of deterministic ones (see, e.g., Ref. [13]), and that any single-party correlation is causal.
- For any given $N \geq 2$ we shall prove the following: if it is true that all $(N-1)$-partite causal correlations can be written as convex combinations of deterministic ones (the induction hypothesis), then the same is true for N-partite causal correlations.

Here goes the proof: Consider an N-partite causal correlation $P(\vec{a}|\vec{x})$, decomposed in the form of Eq. (4.1) (the definition of a causal correlation) which we rewrite here

$$P(\vec{a}|\vec{x}) = \sum_{k \in \mathcal{N}} q_k \, P_k(a_k|x_k) \, P_{k,x_k,a_k}(\vec{a}_{\backslash k}|\vec{x}_{\backslash k}). \tag{4.12}$$

The correlation $P_{k,x_k,a_k}(\vec{a}_{\backslash k}|\vec{x}_{\backslash k})$ is an $(N-1)$-partite causal correlation (for all k, x_k, a_k). The induction hypothesis tells us that the latter can be decomposed as a probabilistic mixture of deterministic causal correlations

$$P_{k,x_k,a_k}(\vec{a}_{\backslash k}|\vec{x}_{\backslash k}) = \sum_{\vec{\alpha}_{\backslash k}} q_{k,x_k,a_k}(k, x_k, a_k) P_{\vec{\alpha}_{\backslash k}}^{\text{det}}(\vec{a}_{\backslash k}|\vec{x}_{\backslash k}) \tag{4.13}$$

where the sum is over all functions $\vec{\alpha}_{\backslash k} : \vec{x}_{\backslash k} \mapsto \vec{a}_{\backslash k}$ that define a deterministic causal correlation $P_{\vec{\alpha}_{\backslash k}}^{\text{det}}(\vec{a}_{\backslash k}|\vec{x}_{\backslash k})$. We have not proved the theorem yet, because we need to express $P(\vec{a}|\vec{x})$ as a convex combination with weights that do not depend on the inputs and outputs. We will see that we can remove this dependency (we just transfer it to the deterministic correlation terms) by proving the following lemma.

Lemma 7 *Consider a set of M points Q_m ($m = 1, \ldots, M$) belonging to some linear space, and Z different points $P(z)$ ($z = 1, \ldots, Z$) in their convex hull, written as convex combinations of the extremal points Q_m in the following way:*

$$P(z) = \sum_{m=1}^{M} q_m(z) \, Q_m \,, \tag{4.14}$$

with weights $q_m(z)$ that depend on z (such that, for each z, all $q_m(z) \geq 0$ and $\sum_{m=1}^{M} q_m(z) = 1$).

Then, each point $P(z)$ can also be written as

$$P(z) = \sum_{m_1=1}^{M} \cdots \sum_{m_Z=1}^{M} \tilde{q}_{m_1,\ldots,m_Z}\, Q_{m_z}, \qquad (4.15)$$

where it is now the extremal points Q_{m_z} that depend on z, while the new weights $\tilde{q}_{m_1,\ldots,m_Z} \geq 0$, $\sum_{m_1,\ldots,m_Z} \tilde{q}_{m_1,\ldots,m_Z} = 1$ are fixed. Note that one of the sums above is over the label m_z, which labels the extremal points.

Proof We start from Eq. (4.15) and prove that it is equivalent to Eq. (4.14). The new weights are defined as

$$\tilde{q}_{m_1,\ldots,m_Z} := \prod_{z=1}^{Z} q_{m_z}(z). \qquad (4.16)$$

where it is clear that they do not depend on z, as the product runs through all the indices z. Note that the weights $q_{m_z}(z)$ are the same to the weights $q_m(z)$ of the decomposition (4.14).

Then for a given z we have that,

$$\sum_{\substack{m_1,\ldots,m_{z-1}, \\ m_{z+1},\ldots,m_Z}} \tilde{q}_{m_1,\ldots,m_Z} = q_{m_z}(z), \qquad (4.17)$$

(because $\sum_{m_i=1}^{M} q_{m_i} = 1$), and

$$\sum_{m_1,\ldots,m_Z} \tilde{q}_{m_1,\ldots,m_Z}\, Q_{m_z} = \sum_{m_z} \sum_{\substack{m_1,\ldots,m_{z-1}, \\ m_{z+1},\ldots,m_Z}} \tilde{q}_{m_1,\ldots,m_Z}\, Q_{m_z}$$

$$= \sum_{m_z} q_{m_z}(z)\, Q_{m_z} = P(z), \qquad (4.18)$$

which completes the proof.   □

Coming back to the proof of Theorem 6, we rename the party-input-output variables $(k, x_k, a_k) \equiv z_k = 1, \cdots, Z_k$. Now we apply Lemma 7 to Eq. (4.13) (the $(N-1)$-partite causal correlation which was decomposed into deterministic correlations)

$$P_{k,x_k,a_k}(\vec{a}_{\backslash k}|\vec{x}_{\backslash k}) = \sum_{\vec{\alpha}_{\backslash k}^{1},\ldots,\vec{\alpha}_{\backslash k}^{Z_k}} \tilde{q}_{\vec{\alpha}_{\backslash k}^{1},\ldots,\vec{\alpha}_{\backslash k}^{Z_k}}\, P_{\vec{\alpha}_{\backslash k}^{z_k}}^{\det}(\vec{a}_{\backslash k}|\vec{x}_{\backslash k}) \qquad (4.19)$$

where the correlations $P^{\text{det}}_{\vec{\alpha}^{z_k}_{\backslash k}}(\vec{a}_{\backslash k}|\vec{x}_{\backslash k})$ are taken from the same set as the $P^{\text{det}}_{\vec{\alpha}_{\backslash k}}(\vec{a}_{\backslash k}|\vec{x}_{\backslash k})$'s above, and hence are deterministic and causal.

We are still on the $(N-1)$-partite correlation. Now we need to prove that the N-partite correlation is a mixture of deterministic ones. One more step before we prove it: we mention the single-party case (which of course can have the desired decomposition) and then we 'combine' the single-party and the $(N-1)$-party correlation like we did earlier in our 'combining correlations one after the other' note.

The single-party probability distributions $P_k(\alpha_k|x_k)$ in Eq. (4.1) can also be decomposed as a mixture of deterministic correlations

$$P_k(\alpha_k|x_k) = \sum_{\alpha_k} P^{det}_{\alpha_k}(\alpha_k|x_k). \tag{4.20}$$

Now using Eqs. (4.19) and (4.20), we can write the correlation $P(\vec{a}|\vec{x})$ of the form of Eq. (4.1) as

$$P(\vec{a}|\vec{x}) = \sum_{k \in \mathcal{N}} q_k \sum_{\alpha_k} q'_{\alpha_k} P^{\text{det}}_{\alpha_k}(a_k|x_k) \times \sum_{\vec{\alpha}^1_{\backslash k}, \dots, \vec{\alpha}^{z_k}_{\backslash k}} \tilde{q}_{\vec{\alpha}^1_{\backslash k}, \dots, \vec{\alpha}^{z_k}_{\backslash k}} P^{\text{det}}_{\vec{\alpha}^{z_k}_{\backslash k}}(\vec{a}_{\backslash k}|\vec{x}_{\backslash k})$$

$$= \sum_{\substack{k, \alpha_k, \\ \vec{\alpha}^1_{\backslash k}, \dots, \vec{\alpha}^{z_k}_{\backslash k}}} q_k \, q'_{\alpha_k} \, \tilde{q}_{\vec{\alpha}^1_{\backslash k}, \dots, \vec{\alpha}^{z_k}_{\backslash k}} \, P^{\text{det}}_{\alpha_k}(a_k|x_k) \, P^{\text{det}}_{\vec{\alpha}^{z_k}_{\backslash k}}(\vec{a}_{\backslash k}|\vec{x}_{\backslash k}) .$$

$$\tag{4.21}$$

which is indeed a convex combination of deterministic causal correlations, with weights independent of the inputs and outputs of any party. This completes the proof.   □

4.4.2 Understanding Deterministic Causal Strategies

As we discussed already, a deterministic strategy (or correlation), $P^{\text{det}}_{\vec{\alpha}}(\vec{a}|\vec{x})$, is characterized by a deterministic function $\vec{\alpha}$ of the list of inputs $\vec{x}$, which determines the list of outputs $\vec{a} = \vec{\alpha}(\vec{x})$. Recall that we require $P^{\text{det}}_{\vec{\alpha}}(\vec{a}|\vec{x})$ to be causal as well—not all functions $\vec{\alpha}$ satisfy this. For the correlation to be causal it must be decomposed in the form of Eq. (4.1). However, $P^{\text{det}}_{\vec{\alpha}}(\vec{a}|\vec{x})$ can only take the value 0 or 1 which implies that the weights in that decomposition must also be 0 or 1, and so the decomposition has only one term. That is, each deterministic causal strategy is compatible with one party acting first.

In particular, a deterministic causal strategy describes the following situation: it determines a party A_{k_1} that acts first. This party obtains an output a_{k_1} as a deterministic function of its input x_{k_1} (because $\vec{a} = \vec{\alpha}(\vec{x})$, which specifies the outputs of all parties). For each action of this party A_k (input x_{k_1}, a_{k_1}), the remaining parties must also share a deterministic causal correlation. We repeat the requirements: one party is determined to act first, which we denote now as $A_{k_2(x_{k_1})}$ because the fact that it acts is determined by the input of the previous party x_{k_1} (we omit the dependence on the output a_{k_1} as it is fully determined by the input x_{k_1}). Now the output of that second party $A_{k_2(x_{k_1})}$ is a deterministic function of its input and the input of the first party. We repeat the pattern for the third party that depends on the input of the first and second party, $A_{k_3(x_{k_1}, x_{k_2})}$, and its output is a deterministic function of its own input and the inputs of the previous guys. And so on and so forth.

The conclusion of the above description is that for each given set of inputs $\vec{x}$, it can be viewed as they are being processed in a particular, generally non-unique, causal order (because some parties may be causally independent which makes them compatible with any causal order) $A_{k_1} \prec A_{k_2^{\vec{x}}} \prec \cdots \prec A_{k_N^{\vec{x}}}$, [with $A_{k_2^{\vec{x}}} = A_{k_2}(x_1)$, etc.]. Hence, the deterministic causal correlation can be written as

$$P_{\vec{\alpha}}^{\text{det}}(\vec{a}|\vec{x}) = P(\vec{a}_{k_1}|\vec{x}_{k_1})\, P(a_{k_2^{\vec{x}}}|x_{k_1}, x_{k_2^{\vec{x}}}) \times \cdots \times P(a_{k_{N-1}^{\vec{x}}}|\vec{x}_{\backslash k_N^{\vec{x}}})\, P(a_{k_N^{\vec{x}}}|\vec{x}) \quad (4.22)$$

4.4.3 What Happens with Trivial Inputs or Trivial Outputs

Some interesting insights arise when we consider causal correlations for a number of parties, where a subset of them have trivial inputs or trivial outputs. Let us start with trivial outputs: consider the case where one party A_k has a fixed output for all possible inputs—or we can ignore that output and equivalently say that A_k has no output. If we look into Bell-type local correlations for this case, this scenario is equivalent to one where the party A_k is simply ignored. In general, if a single input x_k has a fixed output, or equivalently no output, then the local polytope is equivalent to the one obtained by discarding the input x_k completely [14]. Now what can we say about the causal polytope, when A_k has a fixed output for *all* its inputs?

We write the N-partite correlation as

$$P(\vec{a}|\vec{x}) = P(\vec{a}_{\backslash k}|\vec{x}_{\backslash k}, x_k) = P_{x_k}(\vec{a}_{\backslash k}|\vec{x}_{\backslash k}) \quad (4.23)$$

If $P_{x_k}(\vec{a}_{\backslash k}|\vec{x}_{\backslash k})$ is causal for all x_k, then $P(\vec{a}|\vec{x})$ is of the form of Eq. (4.1), and hence causal. Conversely, if $P(\vec{a}|\vec{x})$ is causal, then according to the property of a causal correlation we discussed in the previous section on ignoring the outputs of some parties, $P_{x_k}(\vec{a}_{\backslash k}|\vec{x}_{\backslash k})$ is also causal for each x_k.

Therefore, we can say that an N-partite correlation is causal if and only if all of the conditional $(N-1)$-partite correlations obtained for each possible input x_k of A_k are causal. This is useful in terms of testing whether a correlation $P(\vec{a}|\vec{x})$ is causal, given that A_k has no output: it suffices to test whether the $(N-1)$-partite correlations are causal for each x_k, and one can always assume that A_k acts first. Note that this is not the case when working at the level of process matrices: a party with a trivial output cannot be thought to act first, and it is definitely not the case that if the remaining parties share a causally separable process matrix, then the whole process matrix is causally separable. And we have already seen such an example: the quantum switch. In that tripartite scheme, the parties A and B perform unitaries (and thus have trivial outputs) and the process matrix remaining if we trace out A or B is causally separable. However, we know that the quantum switch is a tripartite causally nonseparable process matrix, and the probabilities arising are causal (they fall into the causal polytope). This is another paradigm of the different levels of investigation: probabilities vs process matrix, which we have discussed extensively in Chap. 2.

Let us move to the other case, where a party A_k has now a single fixed input (or no input). In this case, from the definition of conditional probabilities, we obtain

$$P(\vec{a}|\vec{x}) = P(\vec{a}_{\setminus k}, a_k|\vec{x}_{\setminus k}) = P(\vec{a}_{\setminus k}|\vec{x}_{\setminus k})\, P_{\vec{a}_{\setminus k}, \vec{x}_{\setminus k}}(a_k). \tag{4.24}$$

If $P(\vec{a}_{\setminus k}|\vec{x}_{\setminus k})$ is causal, then $P(\vec{a}|\vec{x})$ is clearly causal (remember our comment on 'combining causal correlations one after the other'). Conversely, if $P(\vec{a}|\vec{x})$ is causal, according to our comment on 'ignoring the outputs of some parties', then $P(\vec{a}_{\setminus k}|\vec{x}_{\setminus k})$ is also causal.

Hence, the causality of an N-partite correlation is equivalent to the causality of the $(N-1)$-partite correlation obtained by discarding party A_k with a fixed input. This is also in the case of locality of correlations (as defined by the local polytope). We can think of A_k as acting after all the $(N-1)$ parties. This is also true in more general situations, whenever a party A_k cannot signal to any other party or set of parties—we can always think that they act last.

4.5 Polytope Characterization

We are ready now to characterize any multipartite causal polytope for any given scenario. The bipartite case was studied in Ref. [12], hence here we study the simplest tripartite case. However, before we do that, we use a simple bipartite example to explain how exactly we characterize a causal polytope.

4.5.1 The Simple Bipartite Case

We consider the simplest bipartite case: both parties have binary inputs and outputs $A : (a, x = 0, 1)$ and $B : (b, y = 0, 1)$ with the restriction (just to keep it simple) that $x = 0 \Rightarrow a = 0$ and $y = 0 \Rightarrow b = 0$. This tells us that for $x = 1$, in general, $a = f(x, y)$ (obviously a will depend only on x if A is first) and that for $y = 1$, in general, $b = g(x, y)$. Each possible combination of the variables (x, y, a, b) will be an axis on the space in which the causal polytope lives. There are two cases: $A \prec B$ and $B \prec A$. For each case, and for different combinations of (x, y, a, b), the probability $P(a, b|x, y)$ might take or not the value 1. This would correspond for a value 0 or 1 to the corresponding axis. These values make up the coordinates of one

Table 4.1 The first two columns are the possible combinations of x, y and the values for a, b given the inputs and the fact that $B \prec A$

x	y	a	b
0	0	0	0
0	1	0	ξ
1	0	η	0
1	1	η_1	ξ

extremal point of the polytope. We first obtain the polytopes corresponding to each causal order $A \prec B$ and $B \prec A$; the causal polytope will be their convex hull.

$B \prec A$: We state the conditions more explicitly, and define some useful variables

$$a = f(x, y) = \eta_{00} + \eta_{01}x + \eta_{10}y + \eta_{11}xy$$
$$\eta = \eta_{00} + \eta_{01} \ (= a, \text{ for } y = 0 \text{ and } x = 1)$$
$$\eta_1 = \eta_{00} + \eta_{01} + \eta_{10} + \eta_{11} \ (= a, \text{ for } y = 1 \text{ and } x = 1) \qquad (4.25)$$
$$b = g(x, y) = g(y) = \xi_0 + \xi_1 y$$
$$\xi = \xi_0 + \xi_1 \ (= b, \text{ for } y = 1).$$

Then, in order to study the correlations $P(a, b|x, y)$, we first find out all the possible values of the variable $a, b, x, y, \xi, \eta, \eta_1$. Table 4.1 is matrix of the possible combinations of a, b, x, y.

From Table 4.1 we can see all the possible arrangements of the inputs and outputs, given that η, η_1, ξ are binary variables. Recall that each combination of (x, y, a, b) written in Table 4.1 and expanded in Table 4.2 for all possible values of $\eta, \eta_1 \xi$, is an axis in the space where the polytope lives. Then the probability $P(a, b|x, y)$ for each combination of the variables (a, b, x, y), specified above, (for different η, η_1, ξ) will be one point in the polytope space. For example, for $a, b, x, y = 0$, the value of $P(a, b|x, y) = 1$ is the coordinate on that axis, for every value of η, η_1, ξ. Another example, for $x, y, a = 1$ and $b = 0$ the value of $P(a, b|x, y)$ is 1 for $\eta = 1$ and 0 for $\eta = 0$. For every combination of the values of η, η_1, ξ, we obtain a different point: hence we will have 8 different points defining our polytope, on the space defined by the axis (a, b, x, y): there are 9 different axes given the above Table and that the variables η, η_1, ξ are binary.

Table 4.2, starting from the sixth row, and bellow the first line, shows the coordinates of each point of $P(a, b|x, y)$ for each axis corresponding to each of the listed combinations of (x, y, a, b). Hence, for every combination of (η, η_1, ξ) there is one vector specified by the its column below. Each vector corresponds to a vertex of the polytope. The convex hull of these vertices is the polytope compatible with $B \prec A$.

In a similar fashion we obtain the vertices for the polytope compatible with $A \prec B$. There are many different softwares that can analyze polytopes given a description of their vertices. Then one can test various properties: if one inequality is a facet of the

Table 4.2 The first four columns are the possible combinations of x, y, a, b for all possible values of η, η_1, ξ. On the top row of the remaining table, is the different combinations of the values η, η_1, ξ and under them is the value $P(a, b|x, y)$. Each value $P(a, b|x, y)$ is a coordinate of the axis specified by the values of x, y, a, b on its left. Then a single point is defined by the column under each η, η_1, ξ combination, in the space defined by the axes (x, y, a, b)

x	y	a	b	000	001	010	011	100	101	110	111
0	0	0	0	1	1	1	1	1	1	1	1
0	1	0	0	1	0	1	0	1	0	1	0
0	1	0	1	0	1	0	1	0	1	0	1
1	0	0	0	1	1	1	1	0	0	0	0
1	0	1	0	0	0	0	0	1	1	1	1
1	1	0	0	1	0	0	0	1	0	0	0
1	1	0	1	0	1	0	0	0	1	0	0
1	1	1	0	0	0	1	0	0	1	0	0
1	1	1	1	0	0	0	1	0	0	0	1

polytope, what are the facets of the polytope, if it is isomorphic to another polytope, etc. After inserting the vertices of the two polytopes, one can obtain the convex hull of the two, namely the causal polytope.

4.5.2 The Simplest Tripartite Case

Now we are ready to characterize the simplest tripartite causal polytope. We will see that the above technique is the first step, but that we easily come across some difficulties: there are too many combinations of (a, b, c, x, y, z) and too many auxiliary variables. Hence, a different parametrization is taken. Also, to enumerate the vertices we no longer do it by hand but with the use of some computational software.

Description of the situation: The simplest tripartite case is the following: each party A_k has a binary input x_k, a single fixed output for one of the inputs, and a binary output for the other, that is, $x_k = 0 \Rightarrow a_k = 0$ and of course $x_k = 1 \Rightarrow a_k = 0$ or 1. We now rename the parties for conveniency and we call them A, B, C with inputs x, y, z and outputs a, b, c. We will denote the complete tripartite probability distribution by P_{ABC}, i.e. $P_{ABC}(a, b, c|x, y, z) := P(a, b, c|x, y, z)$ and by P_{AB}, P_A, etc., the marginal probability distributions for the parties in the subscript. For example, $P_{AB} = (a, b|x, y, z) := \sum_c P_{ABC}(a, b, c|x, y, z)$. Notice that any marginal distribution retains the dependency on all three inputs.

Obtaining the causal polytope: The vertices of the causal polytope for this scenario can be found in the same way as in the bipartite case described above. Namely we enumerate all the deterministic probability distributions $P_{ABC}(a, b, c|x, y, z)$ compatible with all of the 12 possible causal orders. To see these consider the case where one party acts first: there are two fixed causal orders for the remaining parties, and two dynamical causal orders (where each one depends on the first party, as discussed in the previous section). Hence for each party that acts first, we have four possible causal orders. There are three parties, each of them that can be acting first, hence we have 12 possible causal orders. We find that there are 680 deterministic strategies, corresponding to the vertices of the polytope. 480 of those correspond to a fixed causal order and the remaining 192 require dynamical order to be realized.

The causal polytope is 19-dimensional: this is the minimum number of parameters needed to completely specify any probability $P_{ABC}(a, b, c|x, y, z)$. This is because, for each set of inputs (x, y, z), if n of them are nonzero, then one needs $2^n - 1$ values to completely specify the probabilities for these inputs (the -1 comes from normalization of the probabilities). Hence, the dimension of the problem is $\sum_{n=1}^{3} \binom{3}{n}(2^n - 1) = 19$.

To determine the facets of the polytope, which correspond directly to tight causal inequalities, we need to fix a parametrization of the polytope, which means that a set of axes must be defined. Different definitions of these axes lead to different basis representations of the same polytope. Also different a parametrization are more manageable than others by the different softwares used. In fact, the possibility of the convex hull problem to be solved depended critically on the chosen characterization.

On the chosen parametrization (defined in the related paper [3]) the polytope was found to have 13074 facets, each corresponding to a causal inequality. However, there are equivalence classes either by permuting the parties, or by relabeling their outputs, where an inequality can be obtained from others. There are 305 such equivalence classes, whose complete list is in the Supplemental Material of the related paper [3].

Violating the simplest tripartite inequalities: There is a technique, a 'seesaw' approach, that finds violations of causal inequalities (already used in [12]). First, we describe the problem: we need to find a process matrix W and the quantum operations of the parties, described by their CJ matrices $\{M_{a_k|x_k}^{A_k^I A_k^O}\}$ that would produce probabilities $P(\vec{a}|\vec{x})$ that satisfy a given causal inequality of the form $I(P(\vec{a}|\vec{x})) \geq 0$ (where A_k^I, A_k^O is the input and output system of the party A_k). Their produced probabilities (for our three parties) would be

$$P(\vec{a}|\vec{x}) = P(a, b, c|x, y, z) = \mathrm{Tr}\left[\left(M_{a|x}^{A^I A^O} \otimes M_{b|y}^{B^I B^O} \otimes M_{c|z}^{C^I C^O}\right) \cdot W\right], \quad (4.26)$$

(For details in the process matrix formalism, refer to Chap. 2.) The approach we used to find the process matrix and the instruments for the parties is a series of steps that can eventually be realized iteratively by an algorithm.

The protocol: We start with random instruments for the $(N - 1)$ parties. Then we fix a valid process matrix (chosen randomly), and formulate an SDP that finds the optimal instrument of the Nth party such that it minimizes the produced value $I(P(\vec{a}|\vec{x}))$ (remember that we need to violate the inequality $I(P(\vec{a}|\vec{x})) \geq 0$). In the next run of the program, we fix the optimal instrument and vary the process matrix W (we ask the SDP to find the optimal W that minimizes $I(P(\vec{a}|\vec{x}))$). In the next run, we fix the found W and vary the instrument of the N-th party again. After many iterations, we do the same thing with the rest of the parties. Hence, the algorithm initially selects random instruments and a process matrix, and optimizes each of these elements in turn to find a minimum value of $I(P(\vec{a}|\vec{x}))$. This iterative procedure continues until the algorithm converges to a value of $I(P(\vec{a}|\vec{x}))$. Although this only guarantees to find a local minimum, by repeating the procedure many times with different initial instruments one can obtain a bound (if any) on the optimal violation of the causal inequality in question. The process matrix and instruments found for a given causal inequality are presented in Ref. [3].

4.6   Conclusion

In this chapter, based on Ref. [3], we provided the tools to investigate correlations between inputs and output for a number of parties (in other Chapters referred to as settings and outcomes) in terms of whether they respect causality. The reason why we are interested in this task is because there are powerful mathematical and optimization techniques to provide us with the following: given any number of parties, (a) we can obtain the causal polytope by listing its vertices; (b) using a software we can obtain its facets that correspond to causal inequalities; (c) for a given inequality, using optimization techniques we can obtain a process matrix and the operations of the parties that produce correlations for the parties that violate the inequality.

In particular, following Chap. 2 (Ref. [2]), we have the form of causal correlations. These are generated by local operations embedded in a definite causal order—but remember, dynamical. This means that a party can influence the probabilities that different causal relations of future parties are realized. We prove that causal correlations form a polytope whose vertices correspond to deterministic strategies. This means that every possible strategy that obeys causality can be reduced to (probabilistic mixtures of) deterministic ones—where both outputs of the parties and their causal relations are deterministic function of inputs of parties in their past. This significantly simplifies the problem. Using this result, the set of causal correlations can be conveniently described as a convex polytope whose vertices correspond to deterministic strategies. Using polytope characterization techniques, which we describe for the simple bipartite case, we characterized the simplest tripartite causal polytope. We obtained its families of inequalities and with a 'seesaw' approach we showed that one can obtain a process matrix and the parties' operations to produce correlations the violate a given causal inequality. Although the possibility of realizing experimentally such a scenario is still the mothership, here we have laid down a number of tools one can use to obtain correlations and the explicit circumstances for which causality is violated.

References

1. Oreshkov O, Costa F, Brukner Č (2012) Quantum correlations with no causal order. Nat Commun 3:1092
2. Oreshkov O, Giarmatzi C (2016) Causal and causally separable processes. New J Phys 18:093020
3. Abbott AA, Giarmatzi C, Costa F, Branciard C (2016) Multipartite causal correlations: polytopes and inequalities. Phys Rev A 94:032131
4. Bell JS (1964) On the Einstein-Poldolsky-Rosen paradox. Physics 1:195–200
5. Branciard C, Rosset D, Gisin N, Pironio S (2012) Bilocal versus nonbilocal correlations in entanglement-swapping experiments. Phys Rev A 85:032119
6. Fritz T (2012) Beyond bell's theorem: correlation scenarios. New J Phys 14:103001
7. Chaves R, Luft L, Gross D (2014) Causal structures from entropic information: geometry and novel scenarios. New J Phys 16:043001
8. Fritz T (2015) Beyond bell's theorem ii: scenarios with arbitrary causal structure. Commun Math Phys 341:391–434
9. Henson J, Lal R, Pusey MF (2014) Theory-independent limits on correlations from generalized bayesian networks. New J Phys 16:113043
10. Pienaar J (2016) Which causal scenarios are interesting? https://arxiv.org/abs/1606.07798
11. Araújo M et al (2015) Witnessing causal nonseparability. New J Phys 17:102001
12. Branciard C, Araújo M, Feix A, Costa F, Brukner Č (2016) The simplest causal inequalities and their violation. New J Phys 18:013008
13. Fine A (1982) Hidden variables, joint probability, and the bell inequalities. Phys Rev Lett 48:291
14. Pironio S (2005) Lifting bell inequalities. J Math Phys 46:062112

Chapter 5
Experimental Test of a Classical Causal Model for Quantum Correlations

5.1 Back Story

It all started with the results published in Ref. [1]. They studied the traditional Bell scenario: two distant observers perform experiments on their part of a jointly prepared system. The observation of Bell was that the causal structure in which the experiments are embedded, should impose constraints on the correlations that can arise between the outcomes and the settings of the experiments [2], so that those correlations are compatible with a classical causal model for the settings and outcomes. In particular, there were two constraints: 1. Measurement independence: the choice of settings is independent of the state of the joint system, and 2. Locality: the outcomes of one experimenter cannot be influenced by the settings or outcomes of the other experimenter. These constraints, expressed mathematically in a particular scenario specified by the operations of the parties, yield the Bell-type inequalities. For example, when the parties choose between two settings with two outcomes each, we are led to the Clauser-Horne-Shimony-Holt (CHSH) inequalities. However, we all know that these inequalities are predicted to be violated when the joint system is a bipartite entangled state and the experimenters perform quantum measurements on their part of the state.

In Ref. [1], they posed the following question: since the two conceptual constraints disagree with quantum mechanics, how much do we need to relax one or the other constraint to reach an agreement? For example, to what degree can we relax locality—or else to what degree can we allow the outcomes of one observer to depend on what the other experimenter does—so that the conceptual constraints (or the resulting causal model) agrees with the observed correlations? The answer reported in Ref. [1] was two-fold: first they calculated the minimum causal influence between the outcomes of the parties required so that the correlations violating a CHSH inequality are described by a classical causal model with such a causal link; second, they showed that in a scenario different to the CHSH one, the latter causal model cannot explain the correlations no matter the strength of the causal influence between the parties. Hence, with the central object of investigation being a direct

© Springer Nature Switzerland AG 2019
C. Giarmatzi, *Rethinking Causality in Quantum Mechanics*, Springer Theses,
https://doi.org/10.1007/978-3-030-31930-4_5

causal influence between the outcomes of the parties, it was found that a minimum strength is required to explain the CHSH correlations, and an arbitrary strength is insufficient to explain correlations in a more general scenario (with more settings available to the parties). These results caught our attention and led to this project: an experimental test of both theoretical claims. Our results were published in Ref. [3].

5.2   Introduction

The correlations that arise from measurements on entangled systems have always puzzled physicists. In particular, they appeared to contradict two main concepts about the physical world. The first one is the idea that all physical systems should have objective properties, and in particular that the outcomes of measurements should have well-defined values prior to and independent of their measurement. The second one is the idea that causal influences have to be mediated through some physical system that cannot travel faster than the speed of light. These concepts imply restrictions on the observed correlations, known as *local causality*. Together with the fact that the settings of the experiments in a Bell scenario should be chosen freely (referred to as *measurement independence*), local causality leads to the famous Bell-type inequalities, which are violated by observed correlations. Figure 5.1 shows a typical Bell scenario with variables assigned to main events: state preparation Λ, choice of settings X, Y and outcomes A, B.

A natural way to investigate this incompatibility of our (or Bell's) intuition about the physical world and the observed correlations, is to explore what is wrong with our intuition. In particular, we can explore the extent to which we can relax our intuition (the above two assumptions: local causality and measurement independence) to reach an agreement between the results and our understanding of the results [1, 4–11]. Understanding the results would mean that we have an explicit (causal) mechanism of how the variables involved—the variables Λ, X, Y, A, B depicted in Fig. 5.1— produce the observed correlations. Causal modeling seems to be the ideal tool for this investigation [1, 9], as it is a well-defined framework which can be used for causal inference: to translate correlations into cause-effect relations between variables [12, 13].

Causal inference is achieved through two kinds of data: observational and interventional. Discovering causal relations from observational data alone is difficult in general [14–17]. Observational data refers to correlations between variables that occur 'naturally', without any external mechanism that changes the values of these variables (or if so, they are included in the studied scenario); in other words, it means sheer observation. This can be problematic as there may be possibility for causal influences between certain variables but just happens to not occur naturally and hence cannot be detected by observations alone. This problem has been solved by the concept of intervention: if a variable is a cause for another one, then intervening on the first one would change the statistics of the second one. The data collected with this technique is called interventional. Causal models and the concept of intervention

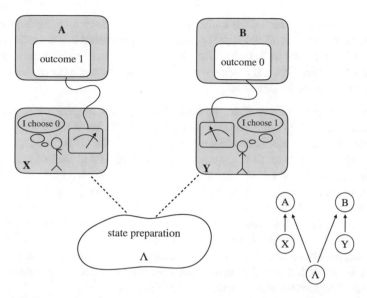

Fig. 5.1 A Bell scenario: where two experimenters receive a part of a jointly prepared state, described by the variable Λ, their choice of setting X and Y and their outcomes A and B. On the bottom right we have the causal model in accordance to the causal assumptions imposed by Bell (locality and measurement independence). As observed correlations are in contradiction with the causal assumptions, this causal model cannot explain the violations of any Bell-type inequality

are the central tools of the theory of causal modeling and using them to formulate our problem provides a clear and quantitative approach to relaxations of the assumptions we are looking at [1, 9]. For example, in a CHSH scenario (which is a Bell scenario where $A, B, X, Y − 0, 1$) we can map the restrictions of Bell's assumptions for the variables into restrictions on the causal model for these variables. Our problem then is the incompatibility of the observed correlations (violating a CHSH inequality) with the resulting causal model derived from Bell's assumptions. This way, when we want to relax any of Bell's assumptions, all we have to do is relax the corresponding constraint on the causal model—it would correspond to the absence or not of a causal link—and compare again the correlations with the new causal model.

Methods: Here, we *test* a class of causal models that relax part of the assumption of local causality. In particular local causality states that there should be no influence from the setting of one party (which we call causal parameter independence) and from the outcome of one party (called outcome independence) to the outcome of the other party. We are interested in the latter. This is because, by relaxing outcome independence alone, the observed correlations can be explained. We test these models in two experimental ways. Firstly, within the CHSH scenario it was found that the causal link under investigation (between the outcomes, say from A to B) should be above a particular value, reported in Ref. [1], and so we investigated the strength of such a possible causal link through interventions: we intervene on one of the variable and observe whether the statistics of the other one change. Secondly, it was found that these models are ruled out by observed correlations in a scenario different to the CHSH, reported in Ref. [1]. Thus, by obtaining correlations that violate this inequality, the class of causal models with a causal link between the outcomes of the experimenters has to be ruled out.

Results: In the first experiment, within the CHSH scenario, by performing controlled interventions on one of the outcomes, we find that the observed change in the statistics of the other outcomes is insufficient: we observed $(2 \pm 2)\%$ change, consistent with no change, in agreement with quantum predictions, and much less than required to explain the observed CHSH violation. In the second experiment, we consider a different scenario to violate a new Bell-type inequality: each party chooses between three measurement settings with two outcomes each. The observed correlations cannot be explained even for arbitrarily strong causal influence between the outcomes of the parties [18]. Although the first method requires detailed knowledge of the physical system under consideration in order to make the desired claim (that we rule out a causal model with a causal influence between the outcomes), the second method is a device-independent way to make the same claim.

5.3 Theory of Causal Modeling

A causal structure for n variables $(V_1, \cdots, V_n)$ is represented by a directed acyclic graph (DAG), an example shown in Fig. 5.2, with nodes representing the variables connected by directed edges representing causal relations [12]. We depict a Bell-scenario, where two parties, Alice and Bob, perform local measurements on their part of a bipartite entangled state. We depict their outcomes with A and B, their settings with X and Y and the description of the input state with a hidden variable Λ (and with small letters their respective values). Now Bell's assumptions of local causality and measurement independence can be formulated in this framework as restrictions on the possible causal models for this scenario. Local causality implies that the outcome of each party should depend only on state of the system and the choice of setting of the respective party, i.e.

$$p(a|x, y, b, \lambda) = p(a|x, \lambda) \tag{5.1}$$

$$p(b|x, y, a, \lambda) = p(b|y, \lambda). \tag{5.2}$$

These constraints can be obtained by restricting the possible causal influences to the outcome of a party, from either the setting or the outcome of the other party. Hence, as we have already said, local causality is the merging of two assumptions: causal parameter independence (no causal link from the measurement setting of a party to the outcome of the other party) and outcome independence (no causal link from one outcome to the other). Measurement independence states that the measurement choices of Alice and Bob, X and Y respectively, are independent of the state of the system. This implies that there is no causal link from Λ, to the variable X or Y; hence

$$p(x, y, \lambda) = \sum_{\lambda} p(x, y) p(\lambda). \qquad (5.3)$$

The causal models compatible with the above assumptions are the well-known Bell-local hidden variable models: $p(a, b|x, y) = \sum_{\lambda} p(a|x, \lambda) p(b|y, \lambda) p(\lambda)$ and are shown in Fig. 5.2. The constraints on the probabilities arising from such a model are known as Bell-type inequalities. In the simplest Bell scenario, where each of the two parties chooses between two settings $(x, y = 0, 1)$ with two possible outcomes each $(a, b = 0, 1)$, any correlations compatible with Bell-local causal models must respect the CHSH inequality

$$S_2 = \langle A_0 B_0 \rangle + \langle A_0 B_1 \rangle + \langle A_1 B_0 \rangle - \langle A_1 B_1 \rangle \leq 2, \qquad (5.4)$$

where $\langle A_x B_y \rangle = \sum_{a,b=0,1} (-1)^{a+b} p(a, b|x, y)$ is the joint expectation value of A_x and B_y.

Aim: We saw that the two assumptions of Bell (local causality and measurement independence) translate to four restrictions on the existence of causal links: 1. no causal link from the setting of a party to the outcome of another party, 2. no causal link between the outcomes of the parties, 3. no causal link between Λ and the setting of each party, and 4. no causal link between the settings of the parties. As we have mentioned, our method of investigation is to see what happens when we relax one of these assumptions, studied theoretically in Ref. [1]. We focus on the second assumption: no causal link from one outcome to the other. We relax this assump-

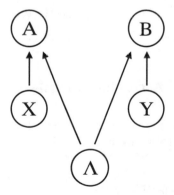

Fig. 5.2 Bell scenario: A Directed Acyclic Graph (DAG), for variables representing the state preparation Λ, settings X, Y and outcomes A, B in a Bell scenario. The causal links between the variables satisfy local causality and measurement independence. The correlations compatible with this model satisfy the CHSH inequality and hence, observed correlations that violate the inequality cannot be described by the depicted causal model—it is a wrong causal model for them

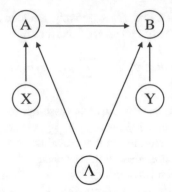

Fig. 5.3 CHSH scenario: A causal model for a Bell scenario, with the additional link from A to B. The insertion of this causal link makes the causal model capable of generating correlations that violate the CHSH inequality. Therefore, this causal model is a possible candidate for the explanation of the observed correlations in a CHSH scenario

tion such that there may be a direct causal influence from say A's outcome to B's outcome, as depicted in Fig. 5.3. The same argument holds for the case of a causal link from B to A or any probabilistic mixture of these cases; this is discussed in the supplementary materials of the related paper [3]. This causal influence could be sub- or super-luminal, instantaneous or even to the past, since the causal model is formulated without any reference to the space-time structure, as long as it does not create any causal loop. However, in our experimental realization, we test the causal link between A and B, where A's events occur in the causal past of B (there is a delay line for the system that reaches B's measurement station. The probability distribution compatible with this causal structure can be decomposed as

$$p(a, b|x, y) = \sum_{\lambda} p(a|x, \lambda) p(b|y, a, \lambda) p(\lambda). \qquad (5.5)$$

5.4 Experiment 1: Interventional Method

The first experiment we perform to test the desired class of causal models (depicted in Fig. 5.3), is an interventional method. This is a core tool in the field of causal discovery that allows us to identify and quantify causal relations [1, 12, 19, 20]. Formally, an intervention is the act of locally forcing a variable X_i to take some specific value x_i' and is denoted as $do(x_i')$. This action breaks all incoming causal arrows to that variable X_i, as the external action is now the only one that defines the value of the variable (see Fig. 5.4), and keeps all other causal arrows untouched. The difficulties in performing such arrow-breaking interventions lies in the fact that it is required to have some background knowledge on the system carrying the causal

**Fig. 5.4 Intervention
model**: The causal model for
our intervention on the
outcome of Alice, A. This
breaks all the incoming
arrows to the variable A

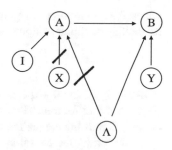

influences; the system on which the measurements occur. This is because there might
be some hidden common causes, which due to our lack of knowledge, we do not
manage to break with our intervention. To imagine this, think of a situation where the
parties share some correlations due to some shared input state using another system
than the ones considered, or the same systems but a different degree of freedom than
the one considered (or even a hidden variable of that system). These correlations
can be used in combination with the measurements to produce the observed correlations. Hence, by breaking all incoming causal arrows with our intervention, does
not exclude any incoming causal arrow using another system or degree of freedom.
This possibility cannot be excluded from the statistics alone, but in our case we shall
regard only the local degrees of freedom and that they behave according to quantum
mechanics. Such assumptions are common in quantum steering scenarios and semi-
device-independent quantum cryptography [21], where the devices of at least one of
the party's measurement stations can be trusted to follow quantum mechanics.

Let us now see what the theoretical predictions are for our experiment. Recall
that our experiment is the usual Bell-scenario, where we allow for the possibility of
a causal link from one outcome to the other. In particular we want to intervene on
one outcome and observe any changes in the statistics of the other. Obviously, there
can be a small or a large change in the statistics; the maximal shift in the probability
distribution of B upon intervention on A permit us to quantify the strength of the
causal link [1]. We use the so-called average causal effect (ACE) [12, 19],

$$\text{ACE}_{A \to B} = \sup_{b,y,a,a'} |p(b|\text{do}(a), y) - p(b|\text{do}(a'), y)|, \qquad (5.6)$$

which is a modification of the measure $C_{A \to B}$ used in Ref. [1]. The latter requires
knowledge about the hidden variable, whereas $\text{ACE}_{A \to B}$ does not and is therefore
experimentally accessible. The average causal effect satisfies the same relation as
$C_{A \to B}$ in Ref. [1], namely,

$$\text{ACE}_{A \to B} \geq \max\left[0, (S_2 - 2)/2\right], \qquad (5.7)$$

where the maximum is taken over all eight symmetries of the CHSH quantity under relabelling of inputs, outputs, and parties [18]. The quantity S_2

$$S_2 = \langle E_{00} \rangle - \langle E_{01} \rangle - \langle E_{10} \rangle + \langle E_{11} \rangle \tag{5.8}$$

(where $\langle E_{a,b} \rangle$, is the expectation value, given the settings of the parties, $\{a, b\} = \{0, 1\}$) and that for the CHSH inequality we have $S_2 \leq 2$. This means that if there is no violation of the inequality, then the $\text{ACE}_{A \to B} = 0$ in order for the correlations to be explained by the causal model—something that makes sense since no violation means that we can explain the correlations anyway. In any other case, the average causal effect has to be larger or equal to a quantity proportional to the CHSH violation achieved by the correlations. This also makes sense as the more 'quantum' the correlations are (the stronger the entanglement, the stronger the violation) the stronger the causal link is required to be to obtain a causal explanation of the correlations (Fig. 5.5).

We experimentally implemented a Bell-scenario with CHSH measurement settings, and performed interventions on the outcome of one station while observing the statistics on the outcomes of the other station. A schematic of the setup is shown in Fig. 5.5. The prepared system is pairs of photons prepared in an entangled state $\cos \gamma |HV\rangle + \sin \gamma |VH\rangle$ in the polarization degree of freedom. H and V correspond to horizontal and vertical polarizations respectively, and γ is the polarization angle of the pump beam, which continuously control the degree of entanglement, as measured by the concurrence, $C = |\sin 2\gamma|$ [22].

The protocol for our two parties, Alice and Bob, is the following: both parties perform their two measurement settings (with two outcomes each) required for the CHSH inequality. The measurement settings measure in the linear-polarization basis ($H - V$ basis), which corresponds to the equatorial plane of the Bloch sphere, see Fig. 5.6. As we mentioned before, we allow for Bob to be in the causal future of Alice, by adding a 2 m fibre delay before Bob's measurement station. The measurement settings for each party is a half-wave plate (HWP) at one of the two usual angles for a maximum violation of the CHSH. Their detection apparatus is a Polarizing Beam Splitter (PBS); each of the two outputs are pointing to an Avalanche Photo Detector (APD) that detect H or V photons. Recall that an intervention on Alice's outcome A needs to break all relevant incoming arrows on that variable and deterministically force the variable to take a particular value. To do this, we cannot simply project the incoming state into any linear combination of H and V because that would break the entanglement of the input state. What we can do though is rotate the state of the incoming photon of Alice to circular polarization states $|R/L\rangle = \frac{1}{\sqrt{2}} (|H\rangle \pm i |V\rangle)$, using a quarter-wave plate (QWP) at $\pm 45°$. This operation, within some experimental precision, erases all relevant information about the setting that Alice used to make her measurement, in the $H - V$ plane. Right after, we project the state in the eigenstates of Alice's PBS, $|H/V\rangle$—which forces one of the two outcomes $A = \pm 1$. This is done with a polarizer (POL) aligned with the PBS. Hence, we can force the outcomes of A to be 1 or -1, independent of what they would have been if the intervention part (rotation to $|R/L\rangle$ states and projection back to $|H/V\rangle$ states) was not there.

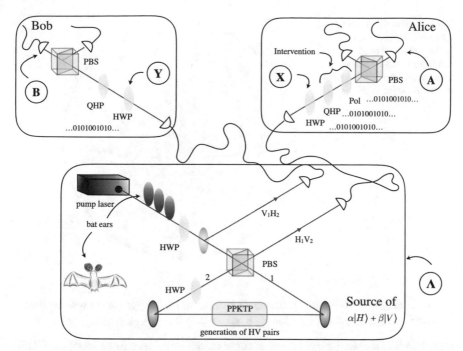

Fig. 5.5 Our experimental setup. The Sagnac source with a periodically poled KTiOPO₄ (ppKTP) crystal: The source of our entangled pairs of photons consists of a pump laser at 405 nm (depicted as purple) followed by 'bat-ear' polarization controllers (a set of three fiber coils which can be rotated around the input fiber's axis) and a HWP which controls the incoming polarization and eventually the degree of entanglement of the source. The horizontal part of the incoming beam goes though the PBS and through the ppKTP where a pair of $|HV\rangle_1$ photons are created (the subscript denotes that the original photon came though the port 1 of the PBS), which turns into a pair of $|VH\rangle_1$ after the HWP in the Sagnac. The H_1 part goes to Alice and the V_1 part to Bob. Now the vertical part of the incoming beam goes in the opposite direction: exits the BPS from port 2, turns into horizontal after the HWP, goes thought the ppKTP where a pair of $|HV\rangle_2$ photos are created, and finally exit the PBS with the V_2 part going to Alice and the H_2 part going to Bob. The pairs of photons created in the ppKTP have half the energy (and frequency, hence double the wavelength, depicted as red) of the incoming photon. The dichroic mirror is transparent on the pump wavelength and reflective on the generated one

This corresponds to operations of the form $|H/V\rangle\langle R/L|$. The measurement bases for Alice and Bob, as well as the setting of the intervention POL and QWP were chosen randomly using quantum random numbers from the Australian National University's online quantum random number generator based on Ref. [23].

Each party owns two APDs, and one of each clicks for each pair of photons. These single-photon clicks are registered with an AIT-TTM8000 time-tagging module. This is an apparatus that attaches a time stamp to each click of an APD, with a temporal resolution of 82 ps. This was done to identify the pairs of photons, or coincidences, arriving in the two measurement stations. An algorithm was written on how to extract this information. Outcome probabilities, used to estimate ACE$_{A \rightarrow B}$, were computed

Fig. 5.6 The Bloch sphere:
where we see the plane on
which the measurements
settings measure and the
operation of the QWP, which
is part of the intervention

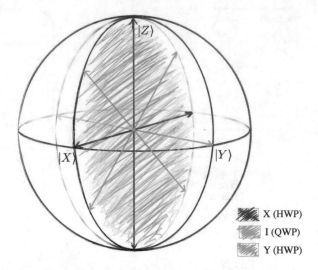

X (HWP)

I (QWP)

Y (HWP)

from a total of 48,000 coincidence counts and no more than one event was registered
for each set of random choices for X, Y, as well as the two elements of I.

Result: We calculated the observed average causal effect as a function of the CHSH
values measured for a range of entangled states (by varying the polarization angle γ of
the pump beam, before the generation of the single-photon pairs). All measured val-
ues are $\mathrm{ACE}_{A \rightarrow B} = 0.02^{+0.02}_{-0.02}$ and largely independent of the observed CHSH violation. The calculated versus the observed average causal effect is depicted in Fig. 5.7.
Note that the theoretical quantity we are trying to observe is bounded from below,
which makes the value zero unachievable in the presence of experimental imperfections and finite counting statistics. Taking this into account, all data lie within 3σ noise
due to Poissonian counting statistics—we expand on this matter on the Supplementary Materials of the related paper [3]. All quoted uncertainties were obtained from
Monte Carlo simulations of the Poissonian counting statistics and correspond to the
0.13th and 99.87th percentile, respectively, which, in the case of normal distributed
variables, would correspond to 3σ confidence regions. Within our experimental capabilities, we found that all CHSH violations above a value of $S_2 = 2.05 \pm 0.02$ cannot
be fully explained with the help of a direct causal link from one outcome to the other.
In other words, the potential causal influence from Alice's outcome to the outcome of
Bob, is not sufficiently strong to account for the observed correlations or the observed
CHSH violation. We remind that quantum mechanics predicts zero causal influence,
a value which we reached within the experimental error.

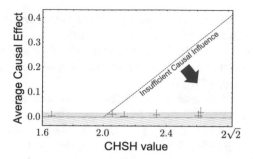

Fig. 5.7 Observed average causal effect: $\text{ACE}_{A \to B}$, for different values of the measured CHSH violation. Any value of $\text{ACE}_{A \to B}$ below the red dashed line, given by Eq. (5.7), is insufficient causal influence to explain the observed CHSH violation. Note that the quantity $\text{ACE}_{A \to B}$ is bounded from below by 0, which explains the asymmetric error distributions. The blue shaded area represents the 3σ region of Poissonian noise. All error bars represent the 3σ statistical confidence obtained from Monte Carlo simulation on the Poissonian counting statistics. From [3]. Reprinted with permission from AAAS

5.5 Experiment 2: Observational Method

Although observational methods for causal inference are widely employed, they do not always provide a conclusive argument to confirm or rule out certain causal models, because they do not yield enough information about the underlying causal model. However, when they do, it is a preferable method because of the strength of the resulting claims on the causal structure. This is because this method is device-independent: no matter what the description of the studied degrees of freedom is, or what the parties' devices are, observation of certain correlations can provide irrefutable constraints on the underlying causal structure. By contrast, as we have seen, interventions can be employed when mere observations fail to provide conclusive answers on the causal structure. In the studied CHSH scenario we saw that interventions were necessary to distinguish direct causation from common-cause correlations. However, this comes at the cost that the intervention relies on the quantum description of the degree of freedom responsible for the outcome A—polarization in our case. Hence the interventional method is device dependent and cannot be employed to test arbitrary hidden-variable models. We show in this section that we can go beyond the CHSH scenario and woulds allow us to perform a device-independent test of any class of models with an arbitrary strong causal influence from one outcome to the other. This will be through the violation of a new Bell-type inequality derived in Ref. [1], which rules out such causal models.

Consider the situation where each of the two parties can choose now between three possible settings, with two outcomes each. As shown in the work of Chaves et al. [1], any correlations compatible with the model in Fig. 5.3 must now satisfy the following inequality

$$S_3 = \langle E_{00} \rangle - \langle E_{02} \rangle - \langle E_{11} \rangle + \langle E_{12} \rangle - \langle E_{20} \rangle + \langle E_{21} \rangle \leq 4. \tag{5.9}$$

The inequality is symmetric under exchange of the parties and it is satisfied by any model compatible with a causal link between the parties' outcomes, or any mixture of them (this is discussed in the Supplementary Materials of the relevant paper [3]). This allows us to test the model in Fig. 5.3 in a device-independent way and irrespective of any temporal ordering of the parties.

To test inequality (5.9), each party performs measurements on their incoming system along one of three directions in the equatorial plane of the Bloch sphere (Fig. 5.6), represented as $O = cos(\theta)Z + sin(\theta)X$. These measurements are implemented using the setup in Fig. 5.5 with the intervention part removed. The specific measurement settings to achieve a maximum violation of the inequality (obtained from numerical results in Ref. [1]) are: $\theta_0^A = -\pi/6$, $\theta_1^A = 7\pi/6$, and $\theta_2^A = -\pi/2$ for Alice and $\theta_0^B = -\pi/3$, $\theta_1^B = \pi/3$, and $\theta_2^B = \pi$ for Bob. Using these measurement settings we obtain

$$S_3 = \frac{3}{2}\sqrt{3}(1 + sin(2\gamma)) \qquad (5.10)$$

Note that since the concurrence is $C = |sin(2\gamma)|$ (which is related to the degree of entanglement), this corresponds to a linear relationship between S_3 and the concurrence of the prepared state. Figure 5.8 shows the observed violation of inequality (5.9) as a function of the parameter γ of the used quantum state. The value of S_3 oscillates between $3\sqrt{3}$ and 0, reaching the maximum for the maximally entangled state at $\gamma = 45°$ (with $C = 1$).

Result: We observed a value of up to $S_3 = 5.16^{+0.02}_{-0.02}$, corresponding to a violation of Eq. (5.9) by more than 170 standard deviations. This observational method completes the interventional method. The latter, rules out outcome-dependent causal models in the CHSH scenario but requires additional assumptions on the underlying causal mechanisms (i.e. that there are no hidden causal mechanisms acting through hidden variables on the examined degree of freedom). However, this observational result rules out outcome-dependent causal models without any additional assumptions, and is valid for any scenario with more than two settings. Hence, in a Bell scenario, a direct causal influence from one outcome to the other cannot explain quantum correlations.

5.6 Conclusion

Our results show that a causal link between the outcomes of two parties in a Bell scenario is found to be insufficient to explain CHSH correlations; ruling out this class of causal models. Furthermore, even if that causal link was there, and is mediated through some experimentally inaccessible hidden variables, we demonstrated that the correlations arising in any more-than-two settings scenario cannot be explained by such a causal influence, no matter how strong. Both experiments rely on the fair-sampling assumption (see Supplementary Materials of the relevant paper [3]

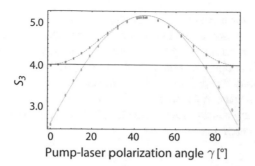

Fig. 5.8 Input state and S_3 value with fixed or non-fixed measurement settings: The orange data points are observed using a fixed measurement scheme, the one that is optimal for the maximally entangled state where $\gamma = 45°$. The orange dotted line represents the corresponding theoretical prediction. The blue data points and blue dashed theory line correspond to the case where the measurement settings were the optimal for the prepared state (see supplementary materials of the relevant paper [3] for the analysis). The black line represents the bound of inequality (5.9): any point above the line cannot be explained by a causal model of the Fig. 5.3. All error bars correspond to 3σ statistical confidence intervals. From [3]. Reprinted with permission from AAAS

for further discussion). Hence, we demonstrated that such causal models—local or nonlocal—are ruled out as candidates to explain quantum correlations.

Besides the fundamental importance of this work—looking for a causal mechanism for quantum correlations—our results could have applications in quantum cryptography scenarios where the secrecy of the measurement outcomes cannot be guaranteed. To see this, consider a one-sided device-independent scenario where Alice's laboratory is trusted, but not Bob's. For example, an eavesdropper, Eve, could control Bob's devices and the source of particles. In a standard quantum key distribution protocol, Alice and Bob would start by certifying that they share entanglement, by performing measurements to violate the CHSH inequality. However, Eve, using her knowledge on the outcomes of Alice (Alice shares them with Bob to calculate the inequality violation, and so Eve receives them since she's controlling— or impersonating—Bob), she can calculate the outcomes she is supposed to obtain to simulate such a violation. Alice can reveal such an attack by performing interventions on her measurement outcome and observing a non-zero value of $ACE_{A \to B}$. Alternatively, Alice and Bob could use inequality (5.9) to certify shared entanglement and prove that there is no Eve. This is because Eve cannot simulate the outcomes of Bob, by knowing the outcomes of Alice.

Our methods of investigation can be used to further explore other classes of causal models, for example, allowing for retrocausal influences or relaxations of measurement independence [1, 6–11, 24]. However, one could take a completely different approach to causality when it comes to quantum systems, the one developed around the process matrix formalism used for quantum causal models [25], or using quantum combs [26]; we will use the former. Quantum causal models are still described by DAGs, with the difference that the nodes do not represent classical variables any more but quantum operations made on quantum systems. In this approach, a quantum causal model for a Bell scenario is shown in Fig. 5.2 with Λ being the input Bell state, X, Y being some input states of Alice and Bob which dictate which setting they will use and A, B being the quantum operations of the parties which yield their outcomes. Although this approach, does not provide a classical explanation to quantum correlations, its applications lie on the exciting new field of quantum causal discovery. It provides the first causal modeling framework that can be used for causal discovery of arbitrary causal structures Ref. [27]. The excited reader is now urged to the next and final chapter.

References

1. Chaves R, Kueng R, Brask JB, Gross D (2015) Unifying framework for relaxations of the causal assumptions in bell's theorem. Phys Rev Lett 114:140403
2. Bell JS (1964) On the Einstein-Podolsky-Rosen paradox. Physics 1:195
3. Ringbauer M et al (2016) Experimental test of nonlocal causality. Sci Adv 2:e1600162
4. Brans CH (1988) Bell's theorem does not eliminate fully causal hidden variables. Int J Theor Phys 27:219–226
5. Branciard C et al (2008) Testing quantum correlations versus single-particle properties within Leggett's model and beyond. Nat Phys 4:681–685
6. Hall MJW (2010) Local deterministic model of singlet state correlations based on relaxing measurement independence. Phys Rev Lett 105:250404
7. Hall MJW (2011) Relaxed bell inequalities and Kochen-Specker theorems. Phys Rev A 84:022102
8. Barrett J, Gisin N (2011) How much measurement independence is needed to demonstrate nonlocality? Phys Rev Lett 106:100406
9. Wood CJ, Spekkens RW (2015) The lesson of causal discovery algorithms for quantum correlations: causal explanations of bell-inequality violations require fine-tuning. New J Phys 17:033002
10. Aktas D et al (2015) Demonstration of quantum nonlocality in the presence of measurement dependence. Phys Rev Lett 114:220404
11. Pütz G, Gisin N Measurement dependent locality (2015). arXiv:1510.09087
12. Pearl J (2009) Causality. Cambridge University Press
13. Spirtes P, Glymour N, Scheines R (2001) Causation, prediction, and search, 2nd edn. The MIT Press
14. Geiger D, Meek C (1999) Quantifier elimination for statistical problems. In: Proceedings of the 15th conference on uncertainty in artificial intelligence, pp 226–235
15. Tian J, Pearl J (2002) On the testable implications of causal models with hidden variables. In: Proceedings of the eighteenth conference on uncertainty in artificial intelligence, pp 519–527. Morgan Kaufmann Publishers Inc
16. Chaves R et al (2014) Inferring latent structures via information inequalities. In: Proceedings of the 30th conference on uncertainty in artificial intelligence, pp 112–121
17. Mooij JM, Peters J, Janzing D, Zscheischler J, Schölkopf B (2014) Distinguishing cause from effect using observational data: methods and benchmarks. arXiv:1412.3773
18. Clauser JF, Horne MA, Shimony A, Holt RA (1969) Proposed experiment to test local hidden-variable theories. Phys Rev Lett 23:880–884
19. Janzing D, Balduzzi D, Grosse-Wentrup M, Schölkopf B (2013) Quantifying causal influences. Ann Statist 41:2324–2358
20. Ried K et al (2015) A quantum advantage for inferring causal structure. Nat Phys 11:414–420
21. Armstrong S et al (2015) Multipartite Einstein–Podolsky–Rosen steering and genuine tripartite entanglement with optical networks. Nat Phys 11:167 EP

22. Hill S, Wootters WK (1997) Entanglement of a pair of quantum bits. Phys Rev Lett 78
23. Symul T, Assad SM, Lam PK (2011) Real time demonstration of high bitrate quantum random
    number generation with coherent laser light. App Phys Lett 98
24. Gallicchio J, Friedman AS, Kaiser DI (2014) Testing bell's inequality with cosmic photons:
    closing the setting-independence loophole. Phys Rev Lett 112:110405
25. Costa F, Shrapnel S (2016) Quantum causal modelling. New J Phys 18:063032
26. Chiribella G, D'Ariano GM, Perinotti P (2008) Transforming quantum operations: quantum
    supermaps 83:30004
27. Giarmatzi C, Costa F (2018) A quantum causal discovery algorithm. npj Quantum Inf 4:17

Chapter 6
A Quantum Causal Discovery Algorithm

6.1 Back Story

After having finished a few projects that further our understanding of the concept
of causality and the process matrix formalism, we were already aware that this
framework can be used for causal discovery with quantum systems, even if the
rigorous ideas were not written yet [1]. By causal discovery, we mean a tool that would
take us from data for a set of variables to the complete causal mechanism (causal
model) that produces the data. We soon realized that the actual steps that would lead
someone from data to a causal model, although rather simple, were numerous and
widely varying from case to case. We could have simply written the most general
instructions on how one can accomplish this task, or, even more exciting, write an
algorithm—the first of its kind—that would be available to anyone. So this was the
task we took upon: write a quantum causal discovery algorithm. The task seemed
simple at the time, but soon we saw its complexity. Nonetheless, some interesting
results surfaced on the way: the different levels of causal information the algorithm
outputs, the fact that it detects Markovianity (a problem already worth to solve on its
own), its potential application to current and future research and finally its potential
extension to be suitable to a much wider variety of applications, like discovering
latent variables. Our results are on Ref. [2] and the code is available on [3].

6.2 Introduction

Discovering causal relations lies at the heart of physics. While observing physical
systems, there is always the question 'how did that come about?' and the answer is a
series of events where one causes the other. While this is a natural tool for formulating
physical processes, only recently was this tool rigorously formalized [4, 5]. By
adding to probability theory the concept of interaction with some external system,
it was possible to formally define what is causation: it is correlation with an extra

© Springer Nature Switzerland AG 2019   125
C. Giarmatzi, *Rethinking Causality in Quantum Mechanics*, Springer Theses,
https://doi.org/10.1007/978-3-030-31930-4_6

ingredient—that of intervention. This led to the development of the framework for causal discovery. Its core ingredients are *causal mechanisms* that are responsible for correlations between observed *events*, with the possibility of external *interventions* on the events.

The concept was not new of course. From simple correlations, like rain and wet land, one can understand that rain causes the land to get wet and not the reverse. To get to this conclusion we need statistics: we can make the land wet and see if rain comes; we can simulate rain with a big hose and see if the land gets wet. The reasoning is the following: it is the fact that we can *make* the one event happen (with intervention) or the other and observe the result, that defines causation. Hence, it is the possibility of interventions that provides an empirically well-defined notion of causation, distinct from correlation: an event A is a cause for an event B if an intervention on A results in a change in the observed statistics of B.

The applications of this simple rule is what allows us to do causal discovery. However, increasing the number of variables makes the task difficult. For two variables, say A and B, if they appear correlated, then intervention on A might change the statistics on B or vice versa; two intervention-experiments are required. For three variables A, B and C, all possible causal links should be checked for any pair of variables, but also conditionally on the third variable. For example, to check a link between A and B we proceed as above, with the extra step that for every intervention-experiment, we set the variable C to take a particular value, and repeat the intervention-experiment for every possible value of C. The whole experiment has to be repeated by permuting all the parties. However, as causal links are established some steps can be dismissed, i.e. if a link from A to B was found, there is no need to check a link from B to A; or if C is a common cause for A and B then it is needless to test any link towards C. This is for the case where the parties have a single fixed causal order between them. If they share a probabilistic mixture of causal orders (or even *dynamical* as we discuss later) then more experiments are needed. This problem sounds ideal for a computer. In that case, all we need to do is perform all possible interventions on all variables and input the results in some algorithm that compares the statistics after each intervention and finally outputs a causal model for the variables. Voilà, a causal discovery algorithm.

The output of a causal discovery algorithm is a causal model. A causal model is typically defined as a set of direct-cause relations and a quantitative description of the corresponding causal mechanisms. The causal relations are represented as arrows in a graph and the causal mechanisms are usually described in terms of transition probabilities (Fig. 6.1). The objective of causal discovery algorithms is to infer a causal model based on observational and interventional data. Such algorithms have found countless applications and constitute one of the backbones in the rising field of machine learning.

But what about developing a causal discovery algorithm for quantum systems? In simple quantum experiments, causal relations are typically known and well under control. However, the fast growth of quantum technologies leads inevitably to networks of increasing size and complexity. Appropriate tools to recover causal relations might become necessary for the functioning of large, distributed quantum networks,

Fig. 6.1 Causal relations:
An example of a causal
relation and its
representation in a graph

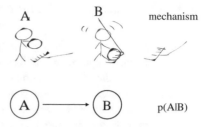

as it is already the case for classical ones [6]. Causal discovery might further detect
the presence of "hidden common causes", namely external sources of correlations
that might introduce systematic errors. From a foundational perspective, the possibility of discovering causal relations from empirically well-defined data opens the
possibility of recovering causal structure from more fundamental primitives.

Classical causal discovery algorithms, however, fail to discover causal relations in
quantum experiments [7]. A considerable effort has been recently devoted to resolve
this tension and transfer causal modeling tools to the quantum domain [8–17]. This
has led to the formulation of a quantum causal modeling framework [1, 18]. (See
Refs. [19, 20] (a paper and a Ph.D. thesis) for a broader philosophical context.)

Here we introduce a first algorithm for the discovery of causal relations in quantum
systems. The starting point of the algorithm is a description of a quantum experiment
(or "process") that makes no prior assumption on the causal relations or temporal
order between events [21]. Given such a description, encoded in a *process matrix*,
the algorithm extracts different levels of causal information about the events in the
experiments. It determines whether or not they are *causally ordered*, namely whether
they can be organised in a sequence where later events cannot influence earlier ones. If
a causal order exists, the algorithm finds whether all common causes are modelled as
events in the process matrix—a property expressed by the condition of *Markovianity*,
as defined in Ref. [1]. If the process is Markovian, the algorithm outputs a causal
model for it: a causal structure (in the form of arrows connecting events) together
with a list of quantum channels and states that generate the process.

The complexity of our algorithm scales quadratically with the number of events,
although the size of the problem itself (the dimension of the process matrix) is
exponential. This suggests that the algorithm can be used efficiently given some
prior assumption that allows an efficient encoding of the input to the code. We
further comment on possible extensions of the algorithm to deal with processes
that are not Markovian, not causally ordered, or that follow different definitions of
Markovianity [18]. We provide the full MatLab code of the algorithm, a Mathematica
file to generate process matrices of random causal structures for testing the code, and
a manual with instructions [3].

6.3 Quantum Causal Models

To do causal discovery, we need data. Classical causal discovery algorithms take typically as input a joint probability distribution for a set of events. This could be simply observational data, or it could be conditioned on external interventions on the events, or the input may be some property of that distribution, like a set of conditional independences. However, when we try to obtain something similar in the quantum case, we immediately run into a problem. The typical description of quantum systems involves the knowledge of a state evolving in time and the temporal ordering of the events is inherently known. For this reason, a different framework must be employed to formulate the problem of quantum causal discovery.

We will use a formulation of quantum mechanics that can assign probabilities to quantum events without any prior knowledge of their causal relations [21]. The process matrix framework is extensively presented in Chap. 2, in Sect. 2.4. Briefly, in this framework, a *quantum event A* is: an experimenter inside a closed laboratory performs some quantum operation on an input system using some quantum instrument and sends out an output system. Formally, it is associated with an input and an output Hilbert space $\mathcal{H}^{A_I}$ and $\mathcal{H}^{A_O}$ respectively—and is represented by a completely positive (CP) map $\mathcal{M}^{A_I \to A_O} : \mathcal{L}(\mathcal{H}^{A_I}) \to \mathcal{L}(\mathcal{H}^{A_O})$, where $\mathcal{L}(\mathcal{H}^S)$ is the space of linear operators over the Hilbert space of system S. A *quantum instrument* is the collection of CP maps $\mathcal{J}^A = \{\mathcal{M}^A\}$, such that $\sum_{\mathcal{M}^A \in \mathcal{J}^A} \mathcal{M}^A$ is a CP and trace-preserving (CPTP) map. $\mathcal{J}$ can be thought to be the choice of operation and the individual map is the associated measurement outcome. For example, a map can be a unitary transformation, a more general CPTP map, or a measurement of the input system followed by a preparation of the output system.

The main result of the process matrix framework is the following: for a set of parties $\{A^1, \cdots, A^n\}$, the joint probability of their CP maps to be realized, given their instruments, is a function of their maps and some matrix that mediates their correlations:

$$p(\mathcal{M}^{A^1}, \cdots, \mathcal{M}^{A^n} | \mathcal{J}^{A^1}, \cdots, \mathcal{J}^{A^n}) =$$
$$\mathrm{Tr}[W^{A_I^1 A_O^1 \cdots A_I^n A_O^n}(M^{A_I^1 A_O^1} \otimes \cdots \otimes M^{A_I^n A_O^n})]. \qquad (6.1)$$

Using a version of the Choi-Jamiołkovsky (CJ) isomorphism [22, 23], the CJ matrix $M^{A_I A_O} \in \mathcal{L}(\mathcal{H}^{A_I} \otimes \mathcal{H}^{A_I})$, isomorphic to a CP map $\mathcal{M}^A : \mathcal{L}(\mathcal{H}^{A_I}) \to \mathcal{L}(\mathcal{H}^{A_O})$ is defined as $M^{A_I A_O} := [\mathcal{I} \otimes \mathcal{M}(|\phi^+\rangle\langle\phi^+|)]^T$, where $\mathcal{I}$ is the identity map, $|\phi^+\rangle = \sum_{j=1}^{d_{A_I}} |jj\rangle \in \mathcal{H}^{A_I} \otimes \mathcal{H}^{A_I}$, $\{|j\rangle\}_{j=1}^{d_{A_I}}$ is an orthonormal basis on $\mathcal{H}^{A_I}$ and T denotes matrix transposition in that basis and some basis of $\mathcal{H}^{A_O}$. Finally, $W^{A_I^1 A_O^1, \cdots, A_I^n A_O^n} \in \mathcal{L}(\mathcal{H}^{A_I^1} \otimes \mathcal{H}^{A_O^1} \otimes \cdots \otimes \mathcal{H}^{A_I^n} \otimes \mathcal{H}^{A_O^n})$ is a matrix that lives on the combined Hilbert space of all input and output systems of the parties and is called *process matrix*.

But what does this framework have to do with causal models? In a causal model, causal relations between events are mediated through some mechanism that allows signaling and is represented by a CPTP map that maps the output of an event to the input of another event. In the process framework, this mechanism corresponds to quantum channels that allow for correlations between the events. These quantum channels are represented by some properties on the process matrix. In particular, we will see later that some properties of the process matrix map perfectly to properties of the underlying causal model. To go back to the very first sentence of this section, that we need data to do causal discovery, we will see that this data helps us obtain the process matrix, and this exactly is the input to the causal discovery algorithm.

In this chapter, we are interested in situations where causal relations define a partial order, which we call *causal order*. We identify causal relations with the possibility of signaling: if the probability of obtaining an outcome in laboratory B can depend on the settings in laboratory A, we say that A *causally precedes* B, and write $A \prec B$. The process matrices that define a causal order between the events are called *causally ordered* and are the subject of our investigation as it is those that encode a causal model. In this Chapter, we will refer only to those process matrices, unless otherwise stated.

6.3.1 Graphical Representation for the Process Framework

The causal structure encoded in the process matrix can be represented by a *Directed Acyclic Graph (DAG)*: A directed graph is a pair $\mathcal{G} = \langle \mathcal{V}, \mathcal{E} \rangle$, where $\mathcal{V} = \{V_1, ..., V_n\}$ is a set of *vertices* (or nodes) and $\mathcal{E} \subset \mathcal{V} \times \mathcal{V}$ is a set of ordered pairs of vertices, representing *directed edges* (Fig. 6.2). A *directed path* is a sequence of directed edges where, for each edge, the second vertex is the first one in the next edge. A *directed cycle* is a directed path that ends up in a vertex already used by the path. A DAG is a directed graph with no directed cycles. As we use a DAG to describe a causal structure, we refer to nodes as parties and to directed edges as causal arrows.

Following Ref. [1], we define a quantum causal model by associating a specific type of process matrix to a DAG. The process matrix framework describes a situation where a number of events, or parties, are correlated through some mechanisms. To

Fig. 6.2 DAG: A typical representation of a Directed Acyclic Graph (DAG)

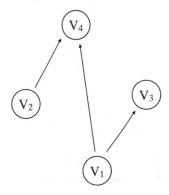

represent this in a DAG, we represent each party with a node in the DAG and their causal relation with an arrow (sometimes referred to as *causal arrows*). Remember that parties are associated with input and output systems and the same now holds for the nodes in the DAG. If the node has more than one outgoing arrow, then the output space is composed of subsystems, with one subsystem for each arrow. We refer to them as *output subsystems*. We define the *parent space* Γ^A of a node A as the tensor product of all output subsystems associated with an arrow ending in A. A *Markov quantum causal model* is then defined by a collection of quantum channels, one for each node A, connecting the parent space of A to its input space.

Now let us see what a process matrix whose causal structure is represented by a DAG looks like. It will be a tensor product of three types of factors: input states for the set of parties with no incoming arrow in the DAG, channels connecting each input system of a remaining party with its parent space, and finally the identity matrix $\mathbb{1}$ for the output systems of the set of parties with no outgoing arrows in the DAG. For example, if $\{F^1, F^2, ..., M^1, M^2, ...L^1, L^2, ...\}$ is a set of parties where F, M and L is the label for the three set of parties described above (first, middle, and last), respectively, then their process matrix would be

$$W^{F_I^1 F_O^1 \cdots} = \rho_1^{F_I^1} \otimes \rho_2^{F_I^2} \otimes \cdots T^{\Gamma^{M^1} M_I^1} \otimes T^{\Gamma^{M^2} M_I^2} \otimes ... \mathbb{1}^{L_O^1 L_O^2 \cdots}. \tag{6.2}$$

The same representation of Markovian processes has recently been used in the study of open quantum systems [24].

The above condition for the causal structure of the process matrix to be described by a DAG is a quantum generalisation of the Markov condition for classical variables and so it can be called the quantum Markov condition [1]. (We will comment below on a slightly different possible definition [18].) Such a process matrix is also causally ordered, with a partial order defined by the DAG. However, the class of causally ordered process matrices is strictly broader than Markovian ones, and they are represented by quantum combs [25].

6.4 Quantum Causal Discovery

Classical causal discovery requires data obtained by performing interventions on the nodes. For quantum causal discovery what is needed is the process matrix W, obtained by (6.1) given the operations of the parties and instruments. So overall, given a W, we can discover the DAG, and with these two we have the underlying quantum causal model.

Before we investigate how to discover the causal model given a process matrix, let us clarify how one obtains that matrix. In Eq. (6.1) we see that the process matrix can be obtained knowing the joint probabilities of all possible maps of a party given their instruments (left-hand side of the equation), and the actual maps of the parties (half of the right-hand side). This is just like quantum state tomography where, using the Born rule, one can reconstruct the quantum state by performing informationally complete measurements on it. In a similar way, by obtaining data where the parties perform informationally complete instruments, one can reconstruct the process matrix [1].

6.4.1 The Linear Constraints

A process matrix of the form of Eq. (6.2) satisfies a set of linear constraints. This set identifies a DAG—in fact, each constraint corresponds to a particular element in the DAG. There are two types of constraints.

Open output: A party A has an *open output* when in the process matrix W there is an identity matrix on the corresponding output system A_O. This translates to the following linear constraint:

$$\tilde{\mathbb{1}}^{A_O} \otimes \text{Tr}_{A_O} W = W \tag{6.3}$$

where $\tilde{\mathbb{1}}^{A_O} = \mathbb{1}^{A_O}/d_{A_O}$ and d_{A_O} is the dimension of the system A_O. When this condition is satisfied, the party A cannot signal to any party and is considered *last*. In the case where the output system of the party is decomposed into subsystems $A_{O_i}, i = 1, \cdots, n$, each corresponding to an outgoing arrow, then the corresponding identity matrix in the process matrix lives on the Hilbert space of that output subsystem A_{O_i}. We also call this subsystem *open* and the linear constraint is

$$\tilde{\mathbb{1}}^{A_{O_i}} \otimes \text{Tr}_{A_{O_i}} W = W \tag{6.4}$$

Channel: A quantum channel between the input of a party A and its parents space Γ^A is represented by a factor $T^{\Gamma^A A_I}$ in the process matrix, as we have already mentioned. It is a positive matrix that lives on the tensor product of the Hilbert spaces of the output and of the input system involved, and has the property that upon tracing out the output of the channel (the input of A) what remains is identity on the input (the space of output systems Γ^A).

$$\text{Tr}_{A_I} T^{\Gamma^A A_I} = \mathbb{1}^{\Gamma^A}. \tag{6.5}$$

This property is necessary and sufficient for the channel to be trace preserving and we use it to discover channels in the process matrix: we trace out the input of A, A_I, and we check whether in the remaining process matrix there is now—and not before—identity on the output system of a given party B. This describes a linear constraint that a process matrix satisfies when there is a channel from the output of B to the input of A

$$\tilde{\mathbb{1}}^{B_O} \otimes \text{Tr}_{B_O}(\text{Tr}_{A_I} W) = \text{Tr}_{A_I} W. \tag{6.6}$$

If the output of party B is decomposed into subsystems, then we use the above constraint for each subsystem separately, by replacing B_O with every output subsystem B_{O_i}

$$\tilde{\mathbb{1}}^{B_{O_i}} \otimes \text{Tr}_{B_{O_i}}(\text{Tr}_{A_I} W) = \text{Tr}_{A_I} W. \tag{6.7}$$

The maximal set of output systems and subsystems for which this condition holds is the parent space of A, Γ^A.

In the concrete implementation of the algorithm, the above equalities are tested up to some precision defined by a small number ϵ, which can be adjusted depending on the working precision. This is due to different numerical rounding of irrational numbers that lead to errors—$\sqrt{2}$ is different to $\sqrt{2}^2/\sqrt{2}$. Naturally, this number can be adjusted to account for experimental inaccuracies.

6.4.2   Ze Code: Theory

The causal discovery code subjects the process matrix to the above types of linear constraints and the set of them that are satisfied define the DAG.

The code takes as input: the number of parties, the dimension of each input system, output system, output subsystem, and the process matrix.

Briefly the procedure of causal discovery can be summarized in three stages. Stage 1: the code identifies the set of parties that are causally independent. This determines if the process matrix is causally ordered. If it is, stage 2 proceeds to trace out any open output subsystems and then to discover the causal arrows between the parties. Stage 3 determines if the process Markovian and if it is, it outputs a DAG. Below we expand on these three stages.

Non-signaling sets for a causally ordered W. Let us call a *non-signaling set*, a set of parties that are causally independent, namely that cannot signal to each other. A non-signalling set is *maximal* if it is not a proper subset of another non-signalling set. The first output of the algorithm is all the maximal non-signalling sets and their causal order. This is done through the linear constraint that detects open output systems, in Eq. (6.3). The set of parties whose output system satisfy the constraint is labeled as *last set*. Note that each constraint has to be satisfied by the whole output system and not by part of it, namely some subsystems. To determine the next set, the *second last*, the code traces out the last set from the process matrix, and using the same constraint it identifies the new last set, and so on. Note that the partition into maximal non-signaling sets does not uniquely identify the partial order of the parties, in the sense that it is not guaranteed that parties in different non-signaling sets can signal to each other. What is guaranteed is that at least one party from a set $\mathcal{X}$ can signal to at least one party in a succeeding set $\mathcal{Y}$ (Fig. 6.3). Note also that the partition into maximal non-signalling sets is not unique, much like a foliation of space-time into space-like hypersurfaces. The way we define this partition is through steps of finding out who is the last party. We are led to a different partition if we go through steps of finding

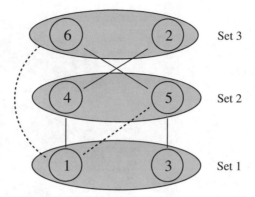

Fig. 6.3 Maximal non-signaling sets: Imagine the output of the code so far is the above groups of parties, representing the non-signaling sets, being the lower one the first set. The actual partial order may be the one represented by the arrows. We can see that although party 1 is a set preceding the set in which party 5 belongs to, this does not mean that there is a causal arrow between them. Same with party 1 and 6

out who is first. In both cases it is still guaranteed that there is a causal arrow from every party in a set $\mathcal{X}$ to at least one party in the succeeding set $\mathcal{Y}$.

The process matrix is causally ordered if and only if the algorithm succeeds in grouping all parties in maximal non-signaling sets. This is because, given the non-signaling sets, we can define a total order among the parties by adding arbitrary order relations among members of each set. For example, we can order the parties in different time steps where when $A \prec B$, A occurs at a time before B, and when $A||B$ then we pick an arbitrary time ordering (Fig. 6.4). With the parties ordered in this way, the process matrix satisfies the condition defining a quantum comb [25]. This is a recursive version of Eq. (6.3), that holds for the output of each system after all systems that come after it are traced out. A central result in the theory of quantum networks is that, whenever this condition holds, the corresponding process can be realised as a channel with memory [25–27]. Thus, this part of the algorithm determines whether the input process matrix has a physical realization as a causally ordered process.

Open output subsystems and causal arrows. The code checks each output subsystem to identify if it is open, using the linear constraint in Eq. (6.4). Each found open subsystem is traced out from the process matrix, keeping track of the label of the party and the label of the subsystem, for example, subsystem 3 of party 2. Keeping track of open subsystems is what allows the algorithm to find a *minimal* DAG, namely without extra arrows, as discussed below.

After the algorithm has traced out all open output subsystems and has established the maximal non-signaling sets of the parties, it is time to determine the DAG. The algorithm is checking all possible causal arrows between pairs of input system of a party and an output system of another party, where each party belongs to a different non-signaling set. For each such pair of parties, say B and A with the former belonging

Fig. 6.4 Total causal order: Starting from the DAG in Fig. 6.3, we can order all events in time, obtaining a total order of the parties, by putting an arbitrary order between parties in the same non-signaling set

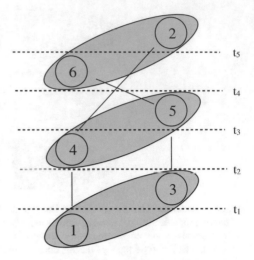

to a set preceding the latter, the code looks at the output system of B and at the input system of A and checks if the linear constraint in Eq. (6.6) is satisfied. If a party's output system is divided into subsystems, then each subsystem is checked using the linear constraint in Eq. (6.7). Every time the constraint is satisfied, an arrow is associated with the corresponding systems and the output system or subsystem is marked as used and is not being checked again. The collection of all output systems and subsystems that satisfy the constraints for a single input system of a party A uniquely identifies the parent space of A, Γ^A. The code first checks pairs of parties that belong to adjacent non-signaling sets, as it is guaranteed to exist at least one causal arrow. Then all the remaining pairs of parties are tested, namely those from distant sets, that still have an unused output system or subsystem.

DAG of a Markovian process. At this stage, the code is ready to output a DAG, for a Markovian process. What remains is to check if it is indeed Markovian, namely if the process matrix is of the form of Eq. (6.2). To do so, the code constructs a test-matrix W_{test} that is Markovian with respect to the found DAG: it contains all (and only) the factors as in Eq. (6.2) that correspond to the elements of the DAG. There are three kinds of these elements: first parties, causal arrows, and last parties; the corresponding terms on the process matrix are input states for the first parties, channels that live on the input and output systems and subsystems of the associated parties, and identity matrices on the output system of the last parties, respectively. To construct the test process matrix, these factors are extracted from the original process matrix by tracing out all systems except the desired ones. If the process is Markovian, then the test-matrix will be equal to the original process matrix that was input to the code.

For example, for a bipartite process, the code may have found a DAG with a causal arrow from A to B. Then it will construct the test-matrix with an input state for A, a channel from A to B and identity for the output system of B. These are extracted from the process matrix, by tracing out all but the desired system, for example, for the input state and the channel, we have the following.

$$\rho^{A_I} = \text{Tr}_{\overline{A_I}} W \tag{6.8}$$
$$T^{A_O B_I} = \text{Tr}_{\overline{A_O B_I}} W$$

where $\overline{A_I}$ means that from the set of all systems involved in the process, this is the subset complementary to the system A_I. The test-matrix would then be $W_{\text{test}} = \rho^{A_I} \otimes T^{A_O B_I} \otimes \mathbb{1}^{B_O}$. If $W = W_{\text{test}}$, then the process matrix is Markovian with respect to the DAG found by the algorithm.

6.4.3 Minimality of the Output DAG

The code is guaranteed to give a unique and minimal DAG for a Markovian process. A process matrix is said to be Markov with respect to the DAG if every channel (found by Eqs. (6.6) and (6.7)) in the process matrix is represented by an arrow in the DAG. However, a W can be Markov to more than one DAG—some DAGs will have arrows allowed by the causal order but there is no actual channel in the W corresponding to this arrow. In other words, a W can be in the tensor product form of Eq. (6.2), but with some factor of the form $T^{\Gamma^M M_I} = \mathbb{1}^{\Gamma^M} \otimes \rho^{M_I}$, for some normalised density matrix ρ. This represents a channel that always produces the state ρ. Hence, this W is Markovian with respect to a DAG with arrows representing such channels, from Γ^M to M_I, but is also Markovian to a DAG without such arrows. Clearly, the same process matrix is still Markovian with respect to a DAG without arrows from Γ^M to M_I.

If every arrow in the DAG corresponds to a non-trivial channel in the process matrix, the DAG is called *minimal*. From another perspective, a DAG is minimal if, by removing any arrow from it, then the W is not any more Markov with respect to the resulting DAG.

The fact that the output of the code is always the minimal DAG is guaranteed by the first step of the algorithm, where the open subsystems are established and discarded. Indeed, an "extra arrow" in a non-minimal DAG would necessarily be associated with an open subsystem—an identity tensor factor in the process matrix. Since the presence of an arrow, condition (6.7), is not checked for open subsystems, the algorithm will not output extra arrows.

Note also that, in [1], it was only proven that a DAG can be in principle recovered under the additional assumption of *faithfulness*. Our algorithm does not require such an extra assumption, proving that causal discovery is always possible for a quantum Markov causal model.

6.5   Complexity of the Algorithm

The dimension of the process matrix is given by the product of input and output dimension of each party. Thus, the size of the process matrix would generally scale exponentially with the number of parties. This is expected, as also the dimension of ordinary density matrices would scale exponentially with the number of parties.

One can however consider situations where, under appropriate assumptions and approximations, the physical scenario under consideration is described by a polynomial number of parameters. Then, the main cost of the algorithm lies in the part that searches for causal arrows between parties. This step tests condition (6.7) for pairs of nodes—the number of tests required is thus quadratic in the number of parties. Therefore, given an efficient encoding of the input process matrix, the algorithm scales polynomially with the number of parties.

6.6   Non-Markovian Processes

A Markovian process is one with a process matrix of the form of Eq. (6.2), and is represented by a DAG. In a non-Markovian process the process matrix is not of that form, i.e. it is not a tensor product of factors representing input states for the parties with no incoming arrows, channels, and identity matrices for the output of the parties in the last set. In other words, in a non-Markovian process, these factors alone—or their representation in a DAG—cannot account for the observed correlations between the events.

6.6.1   Latent Variables

If the code outputs that the process is causally ordered but non-Markovian, it may be the case that there are extra nodes, not considered in the process, which affect the local outcomes of the nodes considered. These are called latent variables [1].

For example, the outcomes of quantum measurements performed in some measurement stations (nodes) in a laboratory, may be affected by the temperature or maybe another system is leaking into one of the stations, like stray light affecting the detection part and causing correlated noise. If these are producing significant change in the data—higher than the noise tolerance in the code—the process will appear non-Markovian.

To recover a causal model by introducing latent variables we would need to extend the algorithm such that it adds nodes and arrows until it finds that is Markovian. Computationally, this task can be hard because the code has to find the right combination of the number of nodes needed, their position in the DAG and the exact channels around them. However, although the original process is non-Markovian, the code still outputs the causal order of the parties for a causally ordered process matrix. From that, one could make guesses about the right causal model, by introducing nodes with specific input and output systems and channels connecting them to the rest of the parties. To do this, one should add the corresponding factors into the current W_{test} and run the code using as input the updated number of parties, dimensions of systems and W_{test} as the process matrix and see if now the process is Markovian.

6.6.2 Mixture of Causal Orders

Another possible reason why the process is non-Markovian is that it might be the case that the process matrix represents a mixture of two or more Markovian processes with different causal order, resulting in a non-causally ordered process matrix.[1] There is a Semidefinite Program (SDP) for this problem, that finds the right decomposition [30]. For instance, for a bipartite process, the SDP would look like the following.

$$
\begin{aligned}
\text{given} \quad & W \\
\text{find} \quad & q \\
\text{such that} \quad & W = q W^{A \prec B} + (1 - q) W^{B \prec A} \\
& 0 \le q \le 0
\end{aligned}
\tag{6.9}
$$

where $W^{X \prec Y}$ denotes a valid process matrix where Y is last and therefore has a factor $\mathbb{1}^{Y_O}$. In the case with more parties, one simply has to write a decomposition that includes all different causal orders for the given parties. Given a decomposition of a process matrix as a mixture of causally ordered ones, one can apply the causal discovery algorithm to each term in the decomposition.

6.6.3 Dynamical and Indefinite Causal Order

So far, we have seen that when events have a definite causal order, they can be represented either by a fixed causal order process or by a mixture of causal orders. However, it may be the case that the process matrix represents a situation of more than two parties, where the causal order of some parties depend on the operations of parties in their past. That is, a party may influence the causal order of future parties. Such a dynamical causal order was studied in [31] where a definition of causality was proposed, compatible with such dynamical causal order. For the tripartite case, it was found that the process matrix describing such a situation should obey certain conditions. However similar conditions were not found for the case of arbitrary

---

[1] A mixture of processes with the same causal order can be modelled as a causally ordered, non-Markovian process with latent nodes acting as "classical common causes" [28, 29].

parties. In such cases, the notion of causal discovery is not clear, as depending on some events in the past, the DAG of future ones would change. Hence the output would be different DAGs for different operations of certain parties. We do not know if the discovery of those DAGs is possible.

6.6.4   Other Definitions of Markovianity

Our algorithm relies on the definition of quantum Markov causal model of Ref. [1]. A different definition was proposed in Ref. [18], where the output systems of the parties are not assumed to factorise into subsystems in the presence of multiple outgoing arrows. In Ref. [18], arrows in the DAG are still associated with a quantum channel from the output space of the parent nodes to the input space of the child but, rather than defining a factorisation in subsystems of the output space, multiple outgoing arrows are more generally associated with commuting channels. For example, in a tripartite scenario where A is a parent of both B and C, a Markovian process matrix would have the form

$$W^{A_I A_O B_I B_O C_I C_O} = \rho^{A_I} \otimes \left(T_1^{A_O B_I} \cdot T_2^{A_O C_I} \right) \otimes \mathbb{1}^{B_O C_O}, \tag{6.10}$$

with the condition $T_1^{A_O B_I} \cdot T_2^{A_O C_I} = T_2^{A_O C_I} \cdot T_1^{A_O B_I}$. Thus, according to Ref. [18], a Markovian process matrix does not need to be a tensor product but can more generally be a product of commuting matrices. To distinguish the two definitions, we will call *tensor-Markovian* and *commuting-Markovian* a process matrix that satisfies the condition of Ref. [1] (used in our code) and Ref. [18], respectively. Note that all tensor-Markovian processes are commuting Markovian, but the converse is not true.[2]

Our algorithm could be adapted to discover the causal structure of commuting-Markovian processes. Note that the strategy used in our code, to detect the parent space of each node by checking (6.5), would not work. Indeed, tracing out B_I from matrix (6.10) does not result in a matrix with identity on A_O. A possible approach could be to instead detect all the *children* of each node A, namely all the nodes with an incoming arrow departing from A. The children are then identified as the smallest subset of parties $C^1, \ldots, C^k$ such that

$$\tilde{\mathbb{1}}^{A_O} \otimes \mathrm{Tr}_{A_O} \left(\mathrm{Tr}_{C_I^1, \ldots, C_I^k} W \right) = \mathrm{Tr}_{C_I^1, \ldots, C_I^k} W. \tag{6.11}$$

---

[2]In Ref. [18] it is further assumed that input and output spaces of each node are isomorphic. Thus, strictly speaking, not all tensor-Markovian processes considered here satisfy the definition of Ref. [18], but only those with input and output of equal dimension. This difference is of little consequence from the point of view of a causal discovery algorithm, since in any case the dimension of each space has to be specified as input to the code.

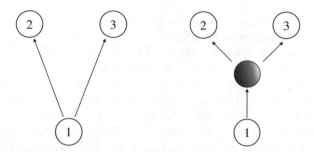

Fig. 6.5 Different definitions of Markovianity: A process that is Markovian according to Ref. [18], e.g. for the DAG on the left, is generally described by a DAG with a latent variable (filled node in the DAG on the right) according to the definition of Markovianity of Ref. [1] on which our algorithm is based

So the code would have to find which parties have to be chosen such that when their input systems are traced out, it leads to identity on the output of party A. As this condition must be checked for subsets of parties, the number of tests is exponential in the number of parties for the worst-case scenario. In contrast, we have seen that to discover a tensor-Markovian causal structure a quadratic number of tests is sufficient. Another potential complication is that our test for Markovianity relies on the tensor-product form of the process matrices; it is not clear if there is a simple way to test whether a process is commuting-Markovian.

An alternative approach is to retain the definition of tensor-Markovian processes and model commuting-Markovian processes as non-Markovian ones. Indeed, since a commuting-Markovian process is causally ordered, it can always be recovered from a tensor-Markovian one by tracing out an appropriate number of latent nodes [1]. An extension of our code to detect latent nodes could thus be used to detect the causal structure of a commuting-Markovian process. In Fig. 6.5 we show an example of a DAG of a commuting-Markovian process (left) and how that would be represented as a tensor-Markovian (right) with a latent variable.

6.7 Conclusions

We have presented and provided an algorithm that can discover an initially unknown causal structure in a quantum network. The first of its type, it is an important proof of principle: it shows that causal structure has a precise empirical meaning in quantum mechanics. Just as with other physical properties, it can be unknown and discovered. This is of particular significance for foundational approaches where causal structure is seen as emergent from more fundamental primitives. Causal discovery provides the methodology to determine when and how causal structure emerges.

Causal discovery can also have broad applications for protocols based on large and complex quantum networks. Our algorithm is guaranteed to find a minimal causal model for any Markovian process, namely a process in which all causally relevant events are under experimental control, with no extra assumptions; this improves on the results of Ref. [1], where the additional condition of faithfulness was invoked. This may have to do with the comparison of classical and quantum causal discovery. In classical causal discovery, the data may by observational or interventional and faithfulness is required which means that one needs to believe that the data are not generated in a way that some conditional independences are hidden. This is because the data may be incomplete. However, in the quantum case we analyzed here using the process matrix framework the data are complete and fully interventional: they arise from informationally complete interventions made at each node. This results to obtaining true conditional independences and no faithfulness condition is needed. Hence, although our method of quantum causal discovery can be reduced to the classical case for a 'classical' process matrix where the parties' operations are diagonal in some fixed basis, the reverse is not true, because it is unclear to define observational data in our quantum case.

Another important use of our algorithm is to tackle the difficult problem of non-Markovianity. An extensive body of research is currently devoted to the problem of detecting and measuring non-Markovianity [32]. Our algorithm finds a concrete solution to the first problem, namely it allows discovering when some external memory is affecting the correlations in the observed system. Detecting non-Markovianity can also have important practical applications for large quantum networks: the presence of "latent nodes", can indicate a possible source of systematic correlated noise in a process, that might affect the working of a quantum protocol. It can further have applications in cryptography for detecting the presence of an eavesdropper. Note that, for non-Markovian processes, the algorithm still recovers important causal information, namely a causal order of the events.

Finally, our algorithm has promising possible extensions. A natural extension is an algorithm that can make "good guesses" for causal structure in the presence of latent nodes. Promising is also the extension of causal discovery to mixtures of causal order, dynamical and indefinite causal structure.

Appendix: Ze Code: Implementation

What follows is a description of the different steps of the code, with a highlight on the main stages that produce the different levels of information about the underlying causal model: causal order, DAG and Markovianity. In certain cases, the intricacy of the problem will be depicted with examples.

Stage 0: Importing and Checking

The code begins with importing all the inputs: the process matrix, the dimension of input and output systems and information about the existence and dimensions of output subsystems.

It checks if there are systems with one dimension. In that case there would be no factor representing them in the process matrix. This is because, the external functions that the code uses (from Tony Cubitt) do not deal well with one dimensional systems. Hence, for every one-dimensional system, a two dimensional identity matrix is inserted, and it is normalized when it is used as an input system.

After, the normalization of the process matrix is checked and it is normalized if it was initially non-normalized. The code outputs this information in the command window.

Stage 1: Causal Order

Definitions: A *non-signaling set* is a set of parties that are causally independent, namely that cannot signal to each other. A non-signaling set is *maximal* if it is not a proper subset of another non-signaling set.

A party A has an *open output* when in the process matrix W there is an identity matrix on the corresponding output system A_O. When this condition is satisfied, the party A cannot signal to any party and is considered *last*.

A process matrix is *causally ordered* if and only if the algorithm succeeds in grouping all parties in maximal non-signaling sets.

The next task is to establish the maximal non-signaling sets and their causal order. The code checks the following constraint for every party, say A

$$\tilde{\mathbb{1}}^{A_O} \otimes \text{Tr}_{A_O} W = W, \tag{6.12}$$

where $\tilde{\mathbb{1}}^{A_O} = \mathbb{1}^{A_O}/d_{A_O}$ and d_{A_O} is the dimension of the system A_O. This constraint checks for open output systems. It finds the parties that are last, which is the last non-signaling set or 'set_1'. Once it establishes this set, it traces it out from the process matrix (traces out all input and output systems of the parties from this set), and the procedure starts again: find out the new set (the second last or 'set_2') and so on. This is a process implemented in a loop, where each run establishes set_1, set_2, etc, where set_1 is the last set, set_2 is the second last, etc.

The procedure of establishing the sets stops when the number of parties that have been used up in this process reaches the number of total parties. The initial number of runs, or of sets that are established, is set to be N; there could be at most N sets with one party in each set. If this process is finished, and the number of parties used is not N, then the process matrix is not causally ordered. The code outputs a message on the command window and the code stops.

The simplified version of the code is the following, where W_n is the input process matrix where the output of party n is traced out and replaced by the identity matrix of the same dimension. Hence, the condition $W_n = W$ for a party n is Eq. (6.3) for a party A.

> **Result**: The maximal non-signaling sets.
> remaining parties = 1:N;
> counter = 0;
> **for** s =*each set;*
> **do**
> > **for** n = *remaining parties;*
> > **do**
> > > **if** *n is last (using condition* (6.3)*)* **then**
> > > > counter = counter+1;
> > > > **if** *counter* = N **then**
> > > > > break;
> > > >
> > > > **end**
> > >
> > > **end**
> >
> > **end**
>
> **end**

Algorithm 1: How the code establishes the maximal non-signaling sets

Output: The code outputs the first result on the command window: the_sets is a matrix where in the first row are the parties of the set_1 (the last parties), second row are the parties of set_2, and so on.

Comment on the complexity: To establish set_1, the code makes N queries to the process matrix. In the worst case scenario, one party will belong to this set. To establish the set_2, the code makes N-1 queries, and so on. N + (N-1) + (N-2) $+ \cdots \sim N^2$. Hence, the total number of queries during this stage is quadratic to the number of parties.

Stage 2: Causal Arrows

Definitions: A directed graph is a pair $\mathcal{G} = \langle \mathcal{V}, \mathcal{E} \rangle$, where $\mathcal{V} = \{V_1, ..., V_n\}$ is a set of *vertices* (or nodes) and $\mathcal{E} \subset \mathcal{V} \times \mathcal{V}$ is a set of ordered pairs of vertices, representing *directed edges*. A *directed path* is a sequence of directed edges where, for each edge, the second vertex is the first one in the next edge. A *directed cycle* is a directed path that ends up in a vertex already used by the path. A DAG is a directed graph with no directed cycles. We refer to edges as *causal arrows*. If every arrow in the DAG corresponds to a non-trivial channel in the process matrix, the DAG is called *minimal*. From another perspective, a DAG is minimal if, by removing any arrow from it, then the W is not any more Markov with respect to the resulting DAG.

The output of this stage is a minimal DAG.

Remove open output subsystems: The first task is to check whether there are any open output subsystems. For party A, this is done using the linear constraint (6.13).

$$\tilde{\mathbb{1}}^{A_{O_i}} \otimes \mathrm{Tr}_{A_{O_i}} W = W. \tag{6.13}$$

When an open output subsystem has been found, there is a message on the command window: "There are open subsystems: subsystem i of party A of dimension d_{A_O}". The code traces out this subsystem from the process matrix. The information that this system was open is being tracked down so that the right labelling of the subsequent subsystems is kept. For instance, if the 3rd subsystem was open, what was earlier the 4th becomes the 3rd, but each time that it is labelled it will still be addressed to as the 4th.

Primal causal arrows: The next task is to check for a causal arrow between the pair of parties. It first checks for *primary* causal arrows, which we call the arrows between adjacent non-signaling sets. We call *secondary* the causal arrows between parties from distant sets. The reason for this distinction between the causal arrows is because it is guaranteed that there is a primal causal arrow between each pair of adjacent sets, whereas this is not the case for secondary arrows.

The code establishes a causal arrow between two parties using the following linear constraint. During this process, the code looks into a pair of parties, belonging to adjacent sets, say party A and B, with $A \prec B$ (in the code named as n1 and n2, respectively). Looking at the output system of A and the input system of B

$$\tilde{\mathbb{1}}^{B_O} \otimes \mathrm{Tr}_{B_O} (\mathrm{Tr}_{A_I} W) = \mathrm{Tr}_{A_I} W. \tag{6.14}$$

Each time the above constraint is satisfied, then the associated output system will not be checked again. An output system cannot belong to more than one causal arrow, whereas an input system can.

If the output of party B is decomposed into subsystems, then the above constraint is checked for each subsystem separately, by replacing B_O with every output subsystem B_{O_i}. Note, again, that each time the above constraint is satisfied, then the associated output subsystem will not be checked again

$$\tilde{\mathbb{1}}^{B_{O_i}} \otimes \mathrm{Tr}_{B_{O_i}} (\mathrm{Tr}_{A_I} W) = \mathrm{Tr}_{A_I} W. \tag{6.15}$$

For each arrow the code will extract the term corresponding to this arrow, in the process matrix—which will be the channel representing the causal arrow if the process is Markovian. This term will be used at the end of the code, to construct the test-process matrix and compare it with the original, to conclude on Markovianity of the process. This term is extracted by tracing out all systems from the process matrix apart from the systems that are examined: the input of B and output of A (be it a system or a subsystem). We can call this the 'complementary system of $A_O B_I$' and denote it by $\overline{A_O B_I}$

$$T^{A_O B_I} = \mathrm{Tr}_{\overline{A_O B_I}} W. \tag{6.16}$$

Below is a simplified version of the part of the code that removes the open output subsystems and establishes the primal causal arrows.

N_sets is the number of sets, established in the previous section.;

```
for set1 =1:N_sets;
do
    for set2 = set1 + 1;
    do
        for n1 =the parties is set1;
        do
            for n2 = the parties in set2;
            do
                if n2 has output subsystem;
                then
                    for each subsystem n2_k;
                    do
                        if n2_k is open;
                        then
                          | Output on the command window, store this info;
                        end
                    end
                    Checked all subsystems of n2, trace out the open ones;
                    Now check for primal causal arrows;
                    for each subsystem n2_k ;
                    do
                        if there is an arrow from n2_k to n1 (using Equation (6.7));
                        then
                          | Extract the corresponding factor in the process matrix;
                          | store it as subterm{n2,n1,k};
                        end
                    end
                else
                    (If n2 has no subsystems.);
                    if there is an arrow from n2 to n1 (using Equation (6.6));
                    then
                      | Extract the channel factor in W; store it as term{n2,n1};
                    end
                end
            end
        end
    end
end
```

Algorithm 2: How the code removes the output open subsystems and establishes the causal arrows

Secondary causal arrows: Now the code proceeds to the discovery of secondary causal arrows. This process is the same as for the primal causal arrows, with the difference that the pair of parties that are checked for a causal arrow do not belong to adjacent sets but distant ones. Again, only remaining output systems and output subsystems are being checked that have not been associated with a causal arrow yet. Again, after an arrow has been established the associated output will not be checked again.

At this stage, a sanity check occurs. The code looks for unused output subsystems of parties that are not last. This should not occur. A party has either an open output subsystem, which at this stage has been detected already and traced out, or an output subsystem that should have been used up in a causal arrow. If the code finds any, it outputs a message on the command window, suggesting that there must be an error in the code, as this should not have happened.

Next the code establishes the parties that are causally independent: this happens either when they are all causally independent and so there is one set which is the first and also the last, or when a party is last and has also no incoming arrow. In any case, there is a message in the command window for the causally independent parties.

Stage 3: Markovianity

Definitions: Markovian process: If $\{F^1, F^2, ..., M^1, M^2, ...L^1, L^2, ...\}$ is a set of parties where F, M and L is the label for the three set of parties described above (first, middle, and last), respectively, then their process matrix would be

$$W^{F_I^1 \Gamma_0^1 ...} = \rho_1^{F_I^1} \otimes \rho_2^{F_I^2} \otimes \cdots T^{\Gamma^{M^1} M_I^1} \otimes T^{\Gamma^{M^2} M_I^2} \otimes ... \mathbb{1}^{L_0^1 L_0^2 ...}, \qquad (6.17)$$

where $T^{\Gamma^{M^j} M_I^j}$ is a matrix representing a CPTP map $\mathcal{T}$ from Γ^{M^j} to M_I^j via the iso-morphism[3] $T^{\Gamma^{M^j} M_I^j} := \mathcal{I} \otimes \mathcal{T}(|\phi^+\rangle\langle\phi^+|) \in \mathcal{H}^{\Gamma^{M^j}} \otimes \mathcal{H}^{M_I^j}$. From now on we identify a channel with its matrix representation.

This is a rather cumbersome process. Although rather simple to explain to a human, the implementation of the code turned out to be quite lengthy—it takes up half the code. So far the code has established the non-signaling sets and the causal arrows, or else, the DAG. The way to test Markovianity is to construct a test-matrix, W_{test}, compatible with only the elements found in the DAG (parties that are first, all channels, parties that are last). These elements are extracted from the process matrix. The rationale is that, if the process is Markovian, then the process matrix will be a tensor product of factors, corresponding only to the elements found in the DAG. Hence, if one constructs a test-matrix that has input states for the first parties,

---

[3]This isomorphism is the same as the one used to describe the CP maps of the parties, but without transposition.

channels connecting each remaining party to their parent space, and identity matrix for the last parties, this matrix should be equal to the original process matrix.

These factors are extracted from the process matrix by tracing out all the systems except for those in which the causal arrow lives. Input states live on the input space of each first party, identity matrices live on the output space of each last party, and channels live on the input space of a party and its parent space (the collection of output systems and output subsystems with an arrow towards the party).

Some of these factors have already been extracted by the code so far. As mentioned in the previous stage, for each causal arrow, the corresponding factor is stored. However, only when the parent space of this arrow is a single output system, this factor is the channel for that arrow and is thus useful. When the parent space is more than one system, to extract the corresponding channel one has to take into account the whole parent space. This is the first task of this stage. Such a factor is extracted by tracing out all the systems except for those that the channel lives on

$$T^{\Gamma^A A_I} = \text{Tr}_{\overline{\Gamma^A A_I}} W. \tag{6.18}$$

Next, the code extracts the input states for the first parties, and the output matrices for the last parties (which should be the identity matrix).

The test-matrix is almost ready. It contains all the elements whose total Hilbert space covers all the input and output systems of the parties. However, if we call W a single *system* that lives on a tensor product of Hilbert spaces of *subsystems*, where each Hilbert space corresponds to inputs and outputs of parties, then these subsystems have to be *ordered* in the process matrix: for three parties with two output subsystems on the second party it would be $W^{A_I A_O B_I B_{O_1} B_{O_2} C_I C_O}$ and the corresponding subsystems would follow that exact order.

For example, if the found DAG is the one from Fig. 6.6, a test-matrix initially will have all the right factors but not in the right order

$$W_{\text{test}} = T^{B_{O_2} C_{O_1} A_I} \otimes T^{C_{O_2} B_I} \otimes T^{B_{O_1} D_I} \otimes \rho^{C_I} \otimes \mathbb{1}^{A_O} \otimes \mathbb{1}^{D_O}. \tag{6.19}$$

The intricacy of this stage lies in the fact that the test-matrix must have the right order for all the inputs and outputs of the parties, namely $A_I A_O \cdots D_O$. Briefly, the code sorts the subsystems such that each input of a party is followed by the output of that party or by the output subsystems of that party. Next, the whole parties (input and output space) are being sorted. Finally, the output subsystems are also sorted.

Output

For a causally ordered process, the code outputs the maximal non-signaling sets and their causal order. Next it outputs the causal arrows, primal and secondary. For a

**Fig. 6.6 Example of a
DAG**: It consists of four
parties, two of which have
two output subsystems

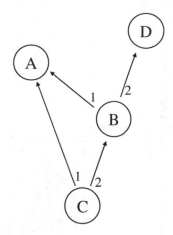

non-Markovian process, these can be ignored. Right after, it outputs if the process is
Markovian or not. For a Markovian process it outputs the DAG.

Below is an example of an output of the code, given some inputs (Table 6.1). The
output of the code is the left part. Comments are added on the right part to enhance
the understanding of each output of the code. These comments are only shown here,
it is not an output of the code. As an example, a Markovian process is used with
parties $\{1, 2, 3, 4\}$ and causal order $\{2 \prec \{3, 4\} \prec 1\}$ and a causal model as shown
in the diagram below.

For the full causal model, one needs the exact mechanisms, namely the channels
and the specific input states of the first parties. These can be accessed as follows.

For the input states, they are stored in the form $\texttt{term\_input[x]}$, where $\texttt{x}$ is a
party that belongs to the first set, i.e. $\{2\}$ in the above example.

For an arrow, in which the receiving party has only one incoming arrow and the
associated output system is not a subsystem, it is stored as $\texttt{term\{x,y\}}$, for the
arrow from party $\texttt{x}$ to party $\texttt{y}$. Also, the last parties are stored like that, with $\texttt{y}$ being
$N + 1$, where N is the total number of parties. In our example there is one term of
that type, which is $\texttt{term\{1,5\}}$ because party 1 is last and $N = 4$. This term is the
identity matrix with the dimension of the output system of that party.

For an arrow, in which the receiving party has only one incoming arrow and the
associated output system is a subsystem, it is stored as $\texttt{subterm\{x, y, z\}}$, for
the arrow from subsystem $\texttt{z}$ of party $\texttt{x}$, to party $\texttt{y}$. In our example it would be:
$\texttt{subterm\{2,4,1\}}$ and $\texttt{subterm\{2,3,2\}}$.

For arrows where the parent space is more than one party, meaning that there
is more than one incoming arrow to the receiving party, the overall channel from
the parent space of party $\texttt{x}$ to party $\texttt{x}$ is stored as $\texttt{subterm\_tot\{x,1\}}$. In our
example, those are $\texttt{subterm\_tot\{1,1\}}$ being a channel from the output of parties
$\{2, 3, 4\}$ to the input of 1 (Fig. 6.7).

Table 6.1 On the left, we depict the output of the code on the command window of MatLab. On the right, we add explain each output information

MatLab output	Explanation
`the_sets =`	*The non-signalling sets, the bottom line is the first set, then the second, etc.*
`    1    0`	
`    3    4`	
`    2    0`	
`Time 0.4487`	*The time it took to evaluate the above.*
	The clock resets here.
	What follows are the primal arrows.
`Link from party 3 to party 1`	*Depicted in the graph.*
`Link from party 4 to party 1`	*Depicted in the graph.*
`Link from subsystem 2 of party 2 to party 3`	*Depicted in the graph without the information on the subsystem 2.*
`Link from subsystem 1 of party 2 to party 4`	*Depicted in the graph without the information 1.*
`4 primal arrows`	*The number of primal arrows*
`primal_arrows =`	*A compact list of the arrows, without any information about the subsystems.*
`    3    1`	*from 3 to 1*
`    4    1`	*from 4 to 1*
`    2    3`	*from 2 to 3*
`    2    4`	*from 2 to 4*
`Time 0.63989`	*The time it took to evaluate the above.*
	Clock resets again.
	What follows is the secondary arrows.
`Link from subsystem 3 of party 2 to party 1`	*Depicted in the graph without the information 3.*
`1 secondary arrows`	*The number of secondary arrows*
`secondary_arrows =`	*A compact list of the arrows.*
`    2    1`	*from 2 to 1*
`Time 0.083953`	*The time it took to evaluate the above.*
`The process is Markovian`	

Fig. 6.7 The code outputs this DAG, given the particular example of a Markovian process matrix

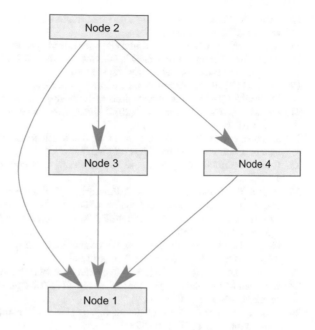

References

1. Costa F, Shrapnel S (2016) Quantum causal modelling. New J Phys 18:063032
2. Giarmatzi C, Costa F (2018) A quantum causal discovery algorithm. npj Quantum Inf 4:17
3. https://github.com/Christina_Giar/quantum-causal-discovery-algo
4. Pearl J (2009) Causality. Cambridge University Press
5. Spirtes P, Glymour CN, Scheines R (2000) Causation, prediction, and search, vol. 81. MIT Press
6. Lamport L (1978) Time, clocks, and the ordering of events in a distributed system. Commun ACM 21:558–565
7. Wood CJ, Spekkens RW (2015) The lesson of causal discovery algorithms for quantum correlations: causal explanations of Bell-inequality violations require fine-tuning. New J Phys 17:033002
8. Tucci RR (1995) Quantum Bayesian nets. Int J Mod Phys B 09:295–337. arXiv:quant-ph/9706039
9. Leifer MS (2006) Quantum dynamics as an analog of conditional probability. Phys Rev A 74:042310
10. Laskey KB (2007) Quantum causal networks. arXiv:0710.1200
11. Leifer MS, Spekkens RW (2013) Towards a formulation of quantum theory as a causally neutral theory of bayesian inference. Phys Rev A 88:052130
12. Cavalcanti EG, Lal R (2014) On modifications of reichenbach's principle of common cause in light of bell's theorem. J Phys A Math Theor 47:424018
13. Fritz T (2015) Beyond bell's theorem II: scenarios with arbitrary causal structure. Commun Math Phys 1–44
14. Henson J, Lal R, Pusey MF (2014) Theory-independent limits on correlations from generalized bayesian networks. New J Phys 16:113043
15. Pienaar J, Brukner Č (2015) A graph-separation theorem for quantum causal models. New J Phys 17:073020

16. Chaves R, Majenz C, Gross D (2015) Information-theoretic implications of quantum causal structures. Nat Commun 6
17. Ried K et al (2015) A quantum advantage for inferring causal structure. Nat Phys 11:414–420
18. Allen JMA, Barrett J, Horsman DC, Lee CM, Spekkens RW (2016) Quantum common causes and quantum causal models. arXiv:1609.09487
19. Shrapnel S (2015) Discovering quantum causal models
20. Shrapnel S (2016) Using interventions to discover quantum causal structure. PhD thesis
21. Oreshkov O, Costa F, Brukner Č (2012) Quantum correlations with no causal order. Nat Commun 3:1092
22. Jamiołkowski A (1972) Linear transformations which preserve trace and positive semidefiniteness of operators. Rep Math Phys 3:275–278
23. Choi M-D (1975) Completely positive linear maps on complex matrices. Linear Algebra Appl 10:285–290
24. Pollock FA, Rodríguez-Rosario C, Frauenheim T, Paternostro M, Modi K (2015) Complete framework for efficient characterisation of non-Markovian processes. arXiv:1512.00589
25. Chiribella G, D'Ariano GM, Perinotti P (2009) Theoretical framework for quantum networks. Phys Rev A 80:022339
26. Gutoski G, Watrous J (2006) Toward a general theory of quantum games. In: Proceedings of 39th ACM STOC, pp 565–574. arXiv:0611234
27. Kretschmann D, Werner RF (2005) Quantum channels with memory. Phys Rev A 72:062323
28. MacLean J-PW, Ried K, Spekkens RW, Resch KJ (2016) Quantum-coherent mixtures of causal relations. arXiv:1606.04523
29. Feix A, Brukner Č (2016) Quantum superpositions of "common-cause" and "direct-cause" causal structures. arXiv:1606.09241
30. Araújo M et al (2015) Witnessing causal nonseparability. New J Phys 17:102001
31. Oreshkov O, Giarmatzi C (2016) Causal and causally separable processes. New J Phys 18:093020
32. Rivas Á, Huelga SF, Plenio MB (2014) Quantum non-Markovianity: characterization, quantification and detection. Rep Prog Phys 77:094001

Chapter 7
Summary

This thesis tackled the concept of causality from two perspectives. First, in a general operational framework where a number of local experiments are described by some probabilistic theory, we asked what are the constraints that causality imposes on their correlations. Our definition of causality, that is, the constraints on the correlations, is compatible with the following intuition: a local experiment cannot affect the occurrence of events in the past or absolute elsewhere of that experiment, nor the strict partial order on such events and the experiment in question. Note that this definition is compatible with a dynamical causal order between the local experiments, that is, one experiment can affect the causal order between future ones. This is an important feature, as no theory of causality has taken such situations into considerations, but only the ones where the causal order is always static.

The second perspective through which we examined causality is in an operational framework where the local experiments are described by standard quantum mechanics. We saw what are the constraints that causality imposes on the possible correlations. This framework was already developed for two parties (local experiments), and we extended it for n parties. The transition from 2 to 3 parties was nontrivial because of the possibility for dynamical causal order we explain above. We found the conditions that causality imposes on the process matrix: the central object of the framework which is used to describe all possible correlations. We commented on the different way in which causality expresses itself within the two different frameworks—with or without local quantum mechanics—and finally studied a few examples of situations that are incompatible with causality. These two perspectives were discussed in Chap. 2. The outcome of this study was a definition of causality compatible with dynamical causal order and the observation that the definition is expressed differently within the two approaches.

Chapter 3 was devoted to the situations that arise from the process matrix framework that are incompatible with causal order, namely their process matrix is 'causally nonseparable'. Given that one such process matrix, the 'quantum switch', has an

© Springer Nature Switzerland AG 2019

C. Giarmatzi, *Rethinking Causality in Quantum Mechanics*, Springer Theses,
https://doi.org/10.1007/978-3-030-31930-4_7

experimental realization, we developed in Chap. 3 a mathematical tool to 'witness' such incompatibility with causality in those realizations. A 'causal witness', similar to an entanglement witness, provides a list of quantum operations that are required to be performed by the parties involved in the realization of the process matrix. Their combined statistics proves causal nonseparability. With the tool of causal witnesses fully developed, one can always check whether a given process matrix is causally separable, and if it is not, it provides the optimal way to witness it in the lab. Note that this is a device-dependent way to detect causal nonseparability.

In Chap. 4, we provided a method to study the correlations compatible with a causal order between the parties, be it fixed, probabilistic or dynamical. In analogy to the local polytope for local correlations, we define the 'causal polytope' where all its points correspond to correlations compatible with causality. The study of a particular causal polytope provides the possible 'causal inequalities' that arise in a given scenario (a number of parties and their settings and outcomes). We provide a way to obtain process matrices that allow for the parties to obtain correlations that violate a given causal inequality. Note that this is a device-independent way of detecting causal nonseparability. Hence, with these computational methods we can study particular scenarios, obtain their causal inequalities and obtain process matrix that violate them. A next step would be to search among the latter, for those that are physically realizable. This would of course be the holy grail, as it is uncertain that such situations can be realized in the lab.

Finally, in Chaps. 5 and 6 we focused on causal inference: assuming the existence of a fixed causal order between a number of events, we aimed to discover their causal relations from the obtained correlations. In Chap. 5 we tested a class of classical hidden-variable causal models for the correlations arising in a Bell scenario. With two complementary experiments we rule out such causal models. In Chap. 6 we delved into a new way of defining causal models for quantum correlations. In this framework, nonsignaling correlations imply causal independence, whereas signaling ones are represented with a causal link between the parties. With this new quantum causal modeling framework at hand, we developed the first algorithm that discovers the underlying causal model of an arbitrary number of parties, given their process matrix and the dimension of their input and output systems. The process matrix can be obtained by performing informationally complete operations at each local experiment. The procedure is essentially that of performing interventions at all the nodes, to discover their causal model. Our algorithm is the first step to causal discovery for quantum systems.

Overall, in this thesis we developed a concept of causality in theory-independent and theory-dependent terms. We saw that the effects of causality are different in the two frameworks. We developed mathematical tools for the theory-dependent case, the process matrix formalism. These helped us probe the interesting cases, i.e those incompatible with causality. First at the level of process matrices, with our causal witnesses, and second at the level of probabilities, with our causal polytopes. We then stepped into the experimental land for a while to rule out a class of causal models that can explain quantum correlations that violate the usual inequality but not another one, and finally we moved to a new causal modeling framework for quantum mechanics.

This time we thought of causality not as an elusive concept that can take indefinite values, neither as a classical concept that we cannot fit into quantum mechanics. Instead, we considered situations with a causal order between a number of local experiments, and we followed a quantum causal modeling framework to develop the first algorithm that, given data from local interventions, it discovers the underlying quantum causal model. That's all folks).

Curriculum Vitae

Dr. Christina Giarmatzi
Quantum Technologies Lab
The University of Queensland
St. Lucia, QLD 4072 Australia

Bio

Christina Giarmatzi is a theoretical quantum physicist from Greece. She specialises in quantum causality and developing computational tools to find answers to foundational questions but also to perform interesting experiments. She currently switched to the study of non-Markovian noise in quantum experiments, and is interested in developing advanced computational and machine learning tools to aid quantum experimentalists dealing with noise, tomography, etc.

She obtained a Degree in Physics, at the Aristotle University of Thessaloniki, in Greece and a Masters in Optics at Institut d'Optique, in Paris. She started a PhD in Brussels which got abandoned to start afresh in Brisbane at The University of Queensland, with two scholarships (IPRS and UQCent). She obtained her PhD with the Dean's Award for outstanding HDR theses. She is currently a Postdoctoral Research Fellow in the same lab of UQ.

Outside the office, she enjoys teaching, mentoring and learning new coding tools for her science.

© Springer Nature Switzerland AG 2019
C. Giarmatzi, *Rethinking Causality in Quantum Mechanics*, Springer Theses,
https://doi.org/10.1007/978-3-030-31930-4

Education

2014–2018 **Ph.D. with Dean's Award** Physics, The University of Queensland,
with Dean's commendation.
2009–2011 **M.Sc.** Optics in Science and Technology, Institut d'Optique, France.
2004–2009 **B.S.** Physics, Aristotle University of Thessaloniki, Greece.

Employment history

2018–present **Postdoctorate Research Fellow** The University of Queensland, Brisbane, Australia
2014–2018 **Ph.D** (obtained) The University of Queensland, Brisbane, Australia
2011–2013 **Ph.D** (not obtained) Université Libre de Bruxelles, Brussels, Bel-
gium
2010–2011 **Intership** (6 months) École Normale Supérieur, Paris, France
2009–2010 **Intership** (3 months) Institut d'Optique, Paris, France

Publications

2018 C. Giarmatzi and F. Costa. Witnessing quantum memory in non-markovian
processes. *arXiv:1811.03722 [quant-ph]*, 2018

K. Goswami, C. Giarmatzi, M. Kewming, F. Costa, C. Branciard, J. Romero,
and A. G. White. Indefinite causal order in a quantum switch. *Physical
Review Letters*, 121(9):090503–08 2018

C. T. M. Ho, F. Costa, C. Giarmatzi, and T. C. Ralph. Violation of a causal
inequality in a spacetime with definite causal order. *arXiv:1804.05498
[quant- ph]*, 2018

2017 C. Giarmatzi and F. Costa. A quantum causal discovery algorithm. *npj Quantum Information*, 4(1):17, 2018

2016 O. Oreshkov and C. Giarmatzi. Causal and causally separable processes. *New
Journal of Physics*, 18(9):093020, 2016

M. Ringbauer, C. Giarmatzi, R. Chaves, F. Costa, A. G. White, and
A. Fedrizzi. Experimental test of nonlocal causality. *Science Advances*, 2(8),
2016

A. A. Abbott, C. Giarmatzi, F. Costa, and C. Branciard. Multipartite causal
correlations: Polytopes and inequalities. *Phys. Rev.* A, 94:032131, Sep 2016

2015 M. Araújo, C. Branciard, F. Costa, A. Feix, C. Giarmatzi, and Č. Brukner. Witnessing causal nonseparability. *New Journal of Physics*, 17(10):102001, 2015

2011 I. H. Agha, C. Giarmatzi, Q. Glorieux, T. Coudreau, P. Grangier, and G. Messin. Time-resolved detection of relative-intensity squeezed nanosecond pulses in an 87 rb vapor. *New Journal of Physics*, 13(4):043030, 2011

Printed in the United States
By Bookmasters